Low-Power CMOS Design for Wireless Transceivers

LOW-POWER CMOS DESIGN
FOR WIRELESS TRANSCEIVERS

ALIREZA ZOLFAGHARI
Resonext Communications, Inc.

Kluwer Academic Publishers
Boston/Dordrecht/London

Distributors for North, Central and South America:
Kluwer Academic Publishers
101 Philip Drive
Assinippi Park
Norwell, Massachusetts 02061 USA
Telephone (781) 871-6600
Fax (781) 681-9045
E-Mail: kluwer@wkap.com

Distributors for all other countries:
Kluwer Academic Publishers Group
Post Office Box 322
3300 AH Dordrecht, THE NETHERLANDS
Telephone 31 786 576 000
Fax 31 786 576 254
E-Mail: services@wkap.nl

 Electronic Services < http://www.wkap.nl>

Library of Congress Cataloging-in-Publication Data
Alireza Zolfaghari
Low-Power CMOS for Wireless Transceivers
ISBN 978-1-4419-5319-3

Printed on acid-free paper.

Printed in the United States of America.

Contents

Contents

List of Figures

List of Tables

List of Tables

Preface

The proliferation of portable communication systems has motivated extensive research on wireless transceivers. In addition to familiar wireless products such as pagers, and cellular phones, many other devices have emerged in the consumer market. Among these, wireless networking shows a great potential to create a fast-growing market.

Because of their popularity and high volumes, cost plays a significant role in wireless devices. Benefiting from the momentum of the digital market, CMOS technology has become attractive to provide low-cost, high-performance transceivers. Furthermore, the portability of wireless devices along with their limited battery life has driven the demand for low-power solutions.

This book deals with design and implementation of low-power wireless transceivers in a standard digital CMOS process. This includes architecture, circuits and monolithic passive components. The low-power techniques are also applied to a 2.4-GHz transceiver, targeting some of the challenging specifications of the two popular standards in this band, i.e., IEEE802.11b and Bluetooth. Since the frequency planning is compatible with GPS signal, a low-IF GPS front end is also accommodated in the prototype.

Chapter 1 introduces wireless networks and describes some of the key features of the two popular standards in the 2.4-GHz ISM band. GPS, which is another fast-growing market in wireless technology, is also briefly discussed in this chapter.

Chapter 2 reviews transceiver architectures, describing the issues with conventional topologies, such as homodyne and heterodyne systems. It then presents the proposed architecture.

Chapter 3 studies stacked inductors and transformers. First, it derives a closed form expression to predict the self-resonance frequency of stacked inductors. The result is then applied to modify these structures. This chapter also includes monolithic transformers with voltage or current gains.

Chapter 4 and 5 concentrate on the design of RF front ends. After discussing conventional low-noise amplifiers and mixers in chapter 4, stacking technique is introduced to save power. This technique is also employed in chapter 5, which describes the transmitter circuits.

Chapter 6 deals with channel-select filters and introduces noninvasive filtering to relax the tradeoffs between power, noise and linearity.

Architecture, circuits and components described in previous chapters are put together in a 2.4-GHz transceiver. Chapter 7 describes the experimental results and chapter 8 presents the conclusions.

This book is based on my Ph.D. work at the University of California, Los Angeles (UCLA). I would like to thank my advisor, Professor Behzad Razavi, for his invaluable guidance, encouragement, and support. I am also grateful to other professors and my friends at UCLA. In particular, I would like to thank Hooman Darabi, Andrew Chan, Jafar Savoj, Lawrence Der, Shahrzad Tadjpour, Shahram Mahdavi, and Shervin Moloudi for helpful discussions.

ALIREZA ZOLFAGHARI

To my parents
and
Anahita

Foreword

As the wireless industry continues to expand to greater dimensions, the demand for new transceiver designs that can reliably operate in a hostile environment rises. Wireless local area networks (WLANs) are a promising domain where low-power, high-performance RF design finds wide application.

The use of CMOS technology for WLAN transceivers is motivated by many recent developments: our understanding of MOS modeling for RF design has improved significantly; passive devices such as inductors have now become common elements in designers' device libraries; and the cost sensitivity of the consumer market has identified CMOS as an important contender.

In this book, Alireza Zolfaghari provides a comprehensive treatment of the challenges in low-power RF CMOS design. He addresses trade-offs and techniques that improve the performance from the component level to the architecture level. He also describes methods of analog and RF design in a standard digital CMOS technology, arriving at a full transceiver solution with a remarkably low power dissipation.

Behzad Razavi
Professor of Electrical Engineering
University of California, Los Angeles

Chapter 1

INTRODUCTION

1.   Wireless Networks

Over recent years, the market for wireless communications has grown rapidly. Wireless technology is now capable of reaching virtually every location on earth.   In addition to familiar wireless products such as pagers, and cellular phones, many other markets have been created. Among these, wireless networks display a great potential for enormous growth.

A wireless local area network (LAN), implemented as an alternative for a wired LAN, provides a flexible data communication system using radio frequency (RF) technology. Eliminating the need for wired connections, wireless LANs transmit and receive data over the air.

The prominent feature of wireless LANs is increased portability. Wireless LANs can provide mobile connectivity in offices, hospitals, factories, etc., allowing users to access real-time information anywhere in their organization.   Another advantage of wireless LAN systems is obviating the need to pull cable through walls and ceilings, increasing installation speed and providing more flexibility. Finally, wireless LAN systems can be employed in a variety of configurations ranging from peer-to-peer networks suitable for a small number of users to full infrastructure topologies of thousands of users.

Similar to wireless LANs, wireless personal area networks (PAN) have also become attractive to provide cordless connections between portable and stationary devices such as personal computers and cell phones. Compared to their LAN counterparts, wireless PANs establish short-range connections at low data rates with little power consumption.

A number of standards have been defined for wireless LAN and PAN systems. Two popular standards in the 2.4-GHz industrial, scientific and medical (ISM) band are IEEE802.11b and Bluetooth. Table 1.1 summarizes some of the key features of these two standards. These standards use different types of spread spectrum, i.e., direct sequence division code multiple access (DS-CDMA) in IEEE802.11b and frequency-hopped spread spectrum (FHSS) in Bluetooth. Since IEEE802.11b uses linear modulation [complementary code keying (CCK)], it requires more linearity in the transmitter.

	IEEE802.11b	Bluetooth
Data Rate	11 Mb/s	1 Mb/s
Multiple Access	DS-CDMA	FHSS
Duplexing	TDD	TDD
Modulation	CCK	GFSK
Sensitivity	-76 dBm	-70 dBm
Bandwidth	22 MHz	1 MHz
Noise Figure	9.6 dB	29 dB
Cost	High	Low

Table 1.1. Wireless standards in the 2.4-GHz band.

Another critical difference between the two standards is the sensitivity. In IEEE802.11b, the sensitivity is −76 dBm across 22 MHz whereas in Bluetooth, the sensitivity is −70 dBm across 1 MHz. Assuming 15 dB SNR for the demodulator, these numbers translate into noise figures of 9.6 dB in IEEE802.11b and 29 dB in Bluetooth. Note that in practice we must target better noise figures to account for the loss of the preselect filter. Finally, since Bluetooth is a cost sensitive standard, we prefer to detect the signal in the analog domain and avoid analog to digital converters.

A number of transceivers have been reported for wireless transceivers in the 2.4-GHz band [2, 3, 4],[5]. Among these, the work in [2] is realized in CMOS technology for Bluetooth and consumes the lowest power (127 mW). For IEEE802.11b which has more stringent specifications, [5] reports 370 mW in a BiCMOS process.

2.   GPS

The global positioning system (GPS) is also another fast-growing market in wireless technology. GPS, which is a constellation of 24 satellites orbiting the earth, provides precise time and position (latitude, longitude and altitude) information on earth. With a GPS receiver, users can determine their location as well as directions. Although originally developed for the military, GPS has proven to be a great asset in a variety of broader civilian, commercial applications.

GPS signals are transmitted from all the satellites at two L-band frequencies: L_1=1575.42 MHz, and L_2=1227.6 MHz. Figure. 1.1 shows how GPS signals are generated. GPS employs binary phase shift keying

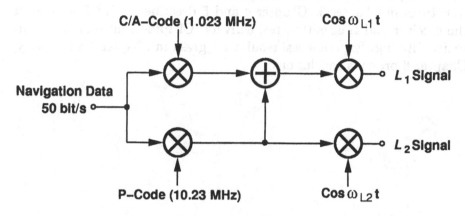

Figure 1.1. GPS signals.

direct sequence spread spectrum (BPSK DSSS). Two codes are used in GPS: the precision code (P code) and the coarse acquisition code (C/A code). Since each satellite is assigned with unique P and C/A codes, the system is referred to as CDMA.

The P-code has a chip rate of 10.23 Mc/s and the C/A code has a chip rate of 1.023 Mc/s. As it is seen, the L_2-band only contains the P code. This band is used to measure the ionospheric delay, providing more precision. However, the P-code is reserved and protected with encryption for military users only. Therefore, commercial GPS receivers are designed for the C/A code, which is available free of charge to all users worldwide.

3. Overview of Topics

The rapid increase in the demand for wireless communications has motivated extensive research on RF transceivers. Due to the limited battery life of wireless products in mobile applications, power consumption becomes a challenging issue for designers. In addition, cost and time to market are also critical factors affecting the choice of the technology. Among different technologies, CMOS, supported by the immense momentum of the digital market, is capable of providing low-cost solutions.

This book describes architecture, circuits and monolithic passive components for low-power wireless transceivers in a standard digital CMOS technology. The low-power techniques are also applied to a 2.4-GHz transceiver targeting some of the challenging specifications of Bluetooth and IEEE802.11b.

Next chapter briefly reviews different architectures and then presents the proposed transceiver architecture. Before presenting circuit building blocks, we study monolithic inductors and transformers with stacked structures in Chapter 3. Chapter 4 and 5 describe the RF front end of the receiver and transmitter, respectively. Chapter 6 introduces noninvasive filtering. Experimental results are given in Chapter 7 and finally, Chapter 8 presents conclusions.

Chapter 2

TRANSCEIVER ARCHITECTURE

1. Introduction

The transceiver architecture is one of the most challenging aspects of the design, impacting complexity, cost, power dissipation, and the number of external components. In this chapter, we describe common receiver architectures including heterodyne, homodyne, and image-reject systems. Next, we review transmitters, focusing on direct-conversion and two-step topologies and finally present the transceiver architecture.

2. Receiver Architectures

A wireless receiver is typically composed of two sections: an analog front end and a baseband digital processor. The analog section receives the modulated radio-frequency (RF) signal and downconverts it to an appropriate intermediate frequency (IF). The downconverted signal can be either demodulated in the analog domain or converted to a digital signal by means of an analog-to-digital converter (ADC) and then demodulated. In order to improve the sensitivity of the receiver, the RF signal is amplified before downconversion and to suppress interferers, the baseband signal is applied to a channel-select filter before detection. RF receivers can be categorized under three principal architectures: heterodyne, homodyne, and image-reject receivers.

2.1. Heterodyne Receivers

Filtering a high-frequency narrow-band signal that is accompanied by large interferers, requires prohibitively high Q's. Therefore, in heterodyne architectures, the signal is downconverted to much lower frequencies so as to relax the required Q. Figure 2.1 shows a simple heterodyne receiver in which the RF signal at ω_{RF} is translated to a lower frequency at $\omega_{IF} = |\omega_{RF} - \omega_{LO}|$. Due to the high noise of the downconversion

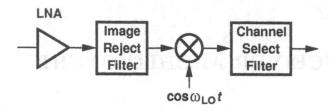

Figure 2.1. Simple heterodyne receiver.

mixer, a low-noise amplifier (LNA) precedes the mixer. The principal consideration here is the image frequency. To understand this issue, note that a simple multiplier does not distinguish between the two inputs at ω_{RF} and $\omega_{image} = |\omega_{RF} - 2\omega_{LO}|$. This is illustrated in Fig. 2.2. The

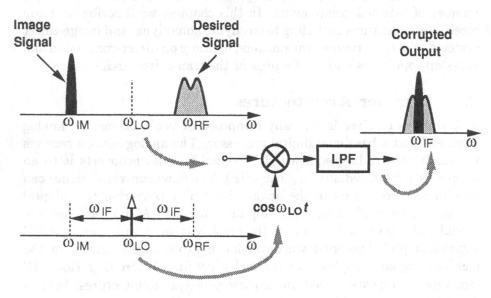

Figure 2.2. Image problem.

most common approach to suppressing the image is through the use of an image-reject filter placed before the mixer. In this architecture, the choice of the IF is very important. A high IF leads to substantial rejection of the image but requires high selectivity in the channel select filter, whereas a low IF requires high selectivity in the image reject filter, inserting more loss in the signal band. Thus, the choice of the IF entails a trade-off between sensitivity and selectivity.

An important drawback of the heterodyne architecture is that the image reject filter is usually realized as a passive, external component. This, furthermore, requires that the LNA drive the 50-Ω input impedance of the filter.

2.2. Homodyne Receivers

Among receiver architectures, homodyne receivers, also called "direct-conversion," look very attractive because of simplicity. As the name implies, the RF signal is directly downconverted to zero-IF by one mixer. Therefore, in this type of the receiver, the LO frequency is equal to the RF. Figure 2.3 shows a simple homodyne receiver.

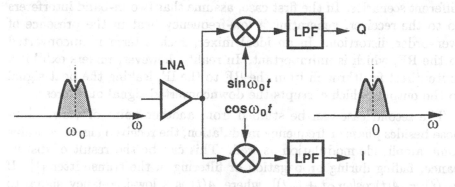

Figure 2.3. Homodyne receiver.

In frequency and phase-modulated signals, the two sides of the spectra carry different information. As a result, for these modulated signals, the downconversion must provide quadrature outputs so as to avoid loss of information.

Due to its simplicity, the homodyne receiver offers two important advantages over a heterodyne counterpart. First, no image filter is required because $\omega_{IF} = 0$. Second, the bandpass IF filter [which is usually an off-chip surface acoustic wave (SAW) filter] is replaced with low-pass filters, which are feasible in monolithic integration.

Although the homodyne architecture is very simple, it has a few drawbacks as follows:

I/Q Mismatch

As explained before, homodyne systems require quadrature mixing. This demands shifting either the RF signal or the LO output by 90°. Since shifting the RF signal generally entails sever noise-power-gain trade-offs, it is more plausible to used the topology in Fig. 2.3. In order to provide quadrature LO signals, one of three approaches can be adopted. First, a quadrature VCO can be used at the cost of higher power dissipation, higher phase noise and greater number of passive components. Second, a divide-by-two circuit can generate quadrature phases if the VCO runs at twice the required frequency. Third, a polyphase filter can be employed to generate the 90° phase shift at the expense of higher power dissipation (due to driver circuits) and more sensitivity to component mismatches.

Heterodyne architectures may also employ I and Q downconversion in the last stage but mismatch requirements are more relaxed because the LO frequency is much lower [1].

Even-Order Distortion

The problem of the second-order distortion can be considered in two different scenarios. In the first case, assume that two in-band interferers go to the receiver, generating a low-frequency beat in the presence of even-order distortion. In an ideal mixer, such a term is upconverted to the RF, which is unimportant. In reality, however, mixers exhibit a finite direct feedthrough from the RF to the IF, leaking the beat signal to the output, which corrupts the downconverted signal of interest.

The second case can be studied from another point of view. Suppose besides phase or frequency modulation, the received signal contains some amplitude modulation as well. This can be the result of disturbance, fading during propagation or filtering in the transmitter [1]. If $x_{in}(t) = A(t)\cos[\omega_{LO}t + \phi(t)]$, where $A(t)$ is a low-frequency signal to model the envelope the AM signal, then second-order distortion yields a term of $A(t)^2/2$. This demodulated AM component can destroy the desired signal.

DC Offset

In a homodyne receiver, the IF is at zero frequency, therefore, offset voltages can corrupt the desired signal and more importantly, saturate the following stages. The offset voltage is caused by the LO leakage to the antenna, LNA, or the input of the mixer. Then the leaked signal is mixed with the LO, producing a DC component. This mechanism is called self mixing. Similarly if a large interferer leaks to the LO port, a dynamic offset is generated at the output of the mixer.

The above discussion suggests that homodyne receivers require some means of offset cancellation. One simple solution to eliminate the DC offset is high-pass filtering [6]. If the spectrum of the signal is not wide enough or if it contains substantial energy near DC, high-pass filtering can substantially destroy the signal. In this case, continuous-time negative feedback or other offset storage techniques must be employed [31].

Flicker Noise

Flicker noise is another source of corruption in the signal band. Since the IF is at zero frequency, the mixer and other stages in the baseband (baseband amplifiers and channel select filters) must be designed so as to minimize the flicker noise. The effect of flicker noise can be reduced by a combination of techniques. In the baseband section, larger devices can be used and in the downconversion stage, mixers with lower flicker noise can be employed. For example, passive mixers have lower flicker noise but they can degrade the noise figure of the system because the gain before the mixer will be limited to that of the LNA.

LO Leakage

Since the LO signal is equal to the RF signal, the LO leakage to the antenna creates interference in the band of other receivers using the same wireless standard.

2.3.   Image-Reject Receivers

Image-reject receivers resolve most of the issues in homodyne systems by two (or more) steps of downconversion. In order to avoid the off-chip image-reject filter in conventional heterodyne systems, they use other image-reject techniques that are more feasible in fully integrated receivers.

All image-reject receivers use quadrature mixers in the first down-conversion and then they eliminate the image signal by different architectures such as Hartly [1], Weaver [1, 28] or polyphase filters [26, 27]. To achieve reasonable image rejection, precise quadrature LO's are required. Furthermore, device and component mismatches substantially degrade the performance of the system. To improve the image rejection of the receive in presence of mismatches and other non-idealities, adaptive calibration methods have been introduced [29, 30]. These techniques, however, increase the complexity of the system.

3.   Transmitter Architectures

An RF transmitter performs modulation, upconversion and power amplification. The modulation can be done in a digital or analog modulator. In some cases, modulation and upconversion are combined in one stage.

Unlike receivers, transmitter architectures are found in only a few forms. This is because issues such as noise, interference rejection and band selectivity are more relaxed in transmitters. Two commonly used architectures are direct-conversion and two-step transmitters.

3.1.   Direct-Conversion Transmitters

Similar to a direct-conversion receiver, if the transmitter carrier frequency is equal to the LO frequency, the architecture is called direct-conversion (Fig. 2.4). Despite simplicity, this architecture suffers from an

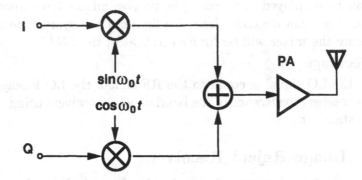

Figure 2.4. Direct-conversion transmitter.

important drawback known as LO pulling. This issue arises when the power amplifier (PA) output which is a high-power modulated signal, corrupts the LO signal.

This phenomenon can be alleviated if the PA output spectrum is sufficiently higher or lower than the oscillator frequency. This can be accomplished by offsetting the LO frequency, i.e., adding or subtracting the output frequency of another oscillator [2], but the LO generator circuit may consume a considerable amount of power and incorporates more spurs.

Another problem with the direct-conversion transmitter is the I/Q mismatch. Since the LO frequency is as high as RF, quadrature LO's are more susceptible to phase mismatches.

3.2.   Two-Step Transmitters

An alternative approach to resolving the problem of LO pulling is to upconvert the baseband signals in two (or more) steps. This is shown in Fig. 2.5. In this case, the PA output spectrum is far from the VCO frequency. Note that a bandpass filter following the second upconversion is required to reject the unwanted sideband by a large factor, typically 50 to 60 dB [1]. This is similar to the image signal in a heterodyne receiver.

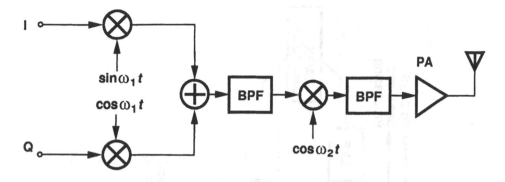

Figure 2.5. Two-step transmitter.

Another advantage of two-step upconversion over direct conversion is that the I and Q matching is more relaxed because the I and Q LO's are at a lower frequency.

4. Proposed Transceiver Architecture

Figure 2.6 shows the transceiver architecture. The receiver employs two downconversion stages using a first LO frequency of 1.6 GHz and a second LO frequency of 800 MHz, translating the input spectrum from 2.4 GHz to an IF of 800 MHz and subsequently to zero. The baseband signals are then applied to either channel-select low-pass filters (LPFs) and a Gaussian frequency shift-keying (GFSK) demodulator (e.g., Bluetooth) or to a baseband processor for direct-sequence spread spectrum (e.g., IEEE802.11b). Similarly, the transmitter upconverts baseband quadrature waveforms (for either linear or nonlinear modulation) to an IF of 800 MHz and afterwards, to 2.4 GHz.

The architecture of Fig. 2.6 has several advantages over typical homodyne or heterodyne counterparts. These advantages are as follows:

1. The LO emission produced by the receiver is well out of the band and heavily suppressed by the selectivity of the antenna.

2. The pulling of the LO by the PA is negligible.

3. Although the system uses two steps of frequency conversion, it requires only one frequency synthesizer.

4. The frequency synthesizer is operating at 1.6 GHz with a channel spacing of 2/3 MHz rather than 1 MHz [32]. Therefore, compared to high-LO architectures, less stringent requirements are placed on the VCO and the prescaler of the synthesizer.

5. An LO frequency of 1.6 GHz allows addition of a low-IF GPS path to the receiver. This will be explained in the next section.

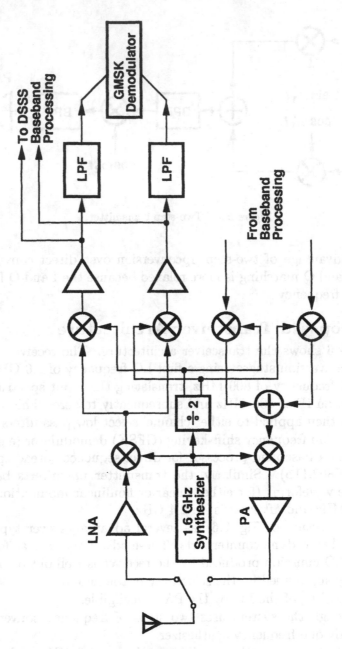

Figure 2.6. Transceiver architecture.

6. Quadrature LO's are generated by a divide-by-two circuit, avoiding power hungry polyphase filters.

7. Since the system is a double-conversion architecture, even-order distortion is not important.

8. Using quadrature mixers operating at 800 MHz improves the I/Q matching in the receiver and transmitter (compared to high-low quadrature mixers used in homodyne and image-reject architectures).

9. Finally, and most importantly, using an LO of 1.6 GHz in the first downconversion reduces the frequency of the image signal to 800 MHz. As a result, unlike conventional heterodyne systems, there is no need for any explicit image-reject filter because as we will show later, the first upconversion stage suppresses the image signal by more than 40 dB. In addition, the antenna and the preselect filter (an off-chip filter that usually precedes the LNA) can provide tens of dB of image rejection. To see how much the antenna can attenuate the image signal, we can refer to Fig. 2.7. This plot shows the measured characteristic of two

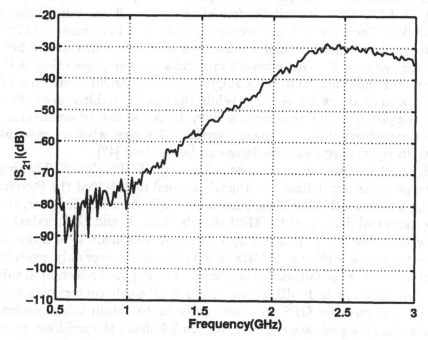

Figure 2.7. Measured characteristic of two antennas operating back to back.

antennas operating back to back. The plot is obtained by measuring the transducer power gain (S_{12}) from one antenna to the other. As the plot shows, each antenna can suppress the image signal by more than 20 dB (Since the total image rejection at 800 MHz is due to two antennas, each antenna attenuates the signal by half of the total rejection at 800 MHz.).

Assuming 40 dB on-chip rejection, the overall image rejection exceeds 60 dB.

The downconversion to zero entails two issues. First, DC offsets must be suppressed, a task feasible by continuous-time negative feedback or other offset storage techniques. Note that in this case, DC offsets are only produced by the LO leakage to the input of the second mixer. Therefore, the DC offset here is lower than that in a homodyne system because the LO leakage to the antenna and LNA and the interferer leakage to the LO port cannot generate any DC offset.

The second issue here is the flicker noise. Nonetheless, the downconversion to zero occurs with an LO frequency of 800 MHz. Thus, the flicker noise of the mixers corrupts the signal to a lesser extent compared to high-frequency quadrature mixers [7].

5.   Compatibility with GPS

As mentioned in the previous chapter, the C/A code of GPS, located at 1.575 GHz, is available free of charge to all commercial users worldwide. There are two architectures widely used in commercial GPS receivers: dual-conversion and single-conversion. Like conventional heterodyne systems, the dual-conversion architecture requires off-chip IF filters. On the other hand, in a single-conversion architecture, the IF can be zero or at low frequencies. While the zero-IF architecture suffers from the problems in direct-conversion systems, the low-IF architecture faces the problem of limited image rejection. However, when we examine the GPS signal spectrum, this issue can be resolved [47].

Figure. 2.8 shows the power spectral density of GPS signals. Looking at the power spectral density of the GPS signal reveals that the P code, having a bandwidth of 20 MHz, surrounds the C/A code. Moreover, the power spectral density of the GPS signals (both P and C/A codes) is below the noise level. As a result, if the C/A code main lobe is downconverted to a low IF (e.g., 2 MHz as in [47]), the receiver only needs to reject the noise of the unwanted sideband. The required rejection in this case can be as low as 10 dB, making the low-IF single-conversion architecture attractive for GPS receivers. In order to obtain this rejection, quadrature IF signals are required. Figure 2.9 shows the architecture of a low-IF GPS receiver.

Using the proposed architecture for 2.4-GHz applications allows us to employ the same LO signal (1.6 GHz) for the receiver of Fig. 2.9, obviating the need for another synthesizer in the GPS receiver.

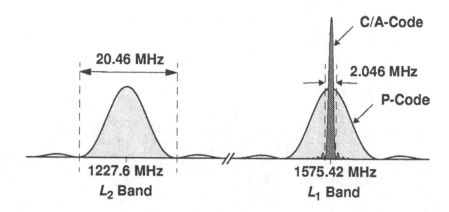

Figure 2.8. Power spectral density of GPS signals.

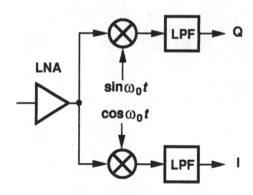

Figure 2.9. Low-IF GPS receiver.

Figure 3.6. Power spectral density of GPS signal

Chapter 3

STACKED INDUCTORS AND TRANSFORMERS

1. Introduction

Monolithic inductors have found extensive usage in RF CMOS circuits. Despite their relatively low quality factor, Q, such inductors still prove useful in providing gain with minimal voltage headroom and operating as resonators in oscillators. Monolithic transformers have also appeared in CMOS technology [8], allowing new circuit configurations.

This chapter introduces a modification of stacked inductors that increases the self-resonance frequency, f_{SR}, by as much as 100%, a result predicted by a closed-form expression that has been developed for f_{SR}. Structures built in several generations of standard digital CMOS technologies exhibit substantial reduction of the parasitic capacitance with the technique applied, achieving self-resonance frequencies exceeding 10 GHz for values as high as 5 nH. The modification allows increasingly larger inductance values or higher self-resonance as the number of metal layers increases in each new generation of the technology.

The chapter also presents a new stacked transformer that achieves nominal voltage or current gains from 2 to 4. Fabricated prototypes display voltage gains as high as 3 in the gigahertz range, encouraging new circuit topologies for low-voltage operation.

The next section of the chapter reviews the definitions of Q. Then we provide the motivation for high-value inductors and summarizes the properties of stacked inductors. Next we deal with the theoretical derivation of the self-resonance frequency of such inductors and Section V exploits the results to propose the modification. Following that we present the stacked transformers and describes a distributed circuit model used to analyze their behavior. Subsequently we summarize the experimental results.

2.   Definitions of the Quality Factor

Several definitions have been proposed for the quality factor. Among these, the most fundamental is:

$$Q = 2\pi \cdot \frac{\text{energy stored}}{\text{energy loss in one oscillation cycle}}. \qquad (3.1)$$

The above definition does not specify what stores or dissipates the energy. However, for an inductor only the energy stored in the magnetic field is of interest. Therefore, the energy stored is equal to the difference between peak magnetic and electric energies.

If an inductor is modeled by a simple parallel RLC tank, it can be shown that [9]

$$\begin{aligned} Q &= 2\pi \cdot \frac{\text{peak magnetic energy} - \text{peak electric energy}}{\text{energy loss in one oscillation cycle}} \\ &= \frac{R_p}{L\omega} \cdot [1 - (\frac{\omega}{\omega_0})^2] \\ &= \frac{Im(Z)}{Re(Z)}, \end{aligned} \qquad (3.2)$$

where R_p and L are the equivalent parallel resistance and inductance, respectively, ω_0 is the resonance frequency and Z is the impedance seen at one terminal of the inductor while the other is grounded. Although definition (3.2) has been extensively used, it is only applicable to the frequencies below the resonance because it falls to zero at the self-resonance frequency.

On the other hand, if only the magnetic energy is considered, then (1) reduces to

$$\begin{aligned} Q &= 2\pi \cdot \frac{\text{peak magnetic energy}}{\text{energy loss in one oscillation cycle}} \\ &= \frac{R_p}{L\omega}. \end{aligned} \qquad (3.3)$$

Definition (3.3) has two advantages over (3.2). First, it can be used over a wider frequency range. Second, it can more explicitly express R_p. It should be noted that at low frequencies, the Q's obtained by (3.2) and (3.3) are quite close because the energy stored in the electric field is much smaller than that stored in the magnetic field.

3. Large inductors with high self-resonance frequencies

Inductors are extensively used in tuned amplifiers and mixers with high intermediate frequencies (IFs) (Fig. 3.1).

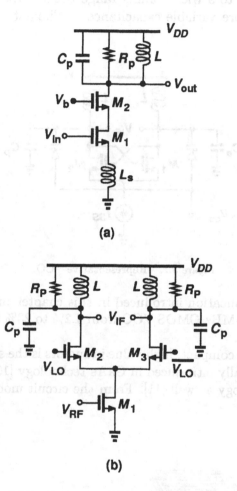

(a)

(b)

Figure 3.1. (a) Low-noise amplifier and (b) mixer with high IF.

In these applications, to maximize the gain (or conversion gain), the equivalent parallel resistance of the inductor (R_p) must be maximized. From definition (3.3) of the Q, R_p can be expressed as

$$R_p = Q \cdot L\omega. \tag{3.4}$$

Therefore, to maximize R_p, the *product* of Q and L must be maximized. Since the Q of on-chip inductors in CMOS technology is quite limited, it is reasonable to seek methods of achieving high inductance values with high self-resonance frequencies and a moderate silicon area.

If a method of reducing the parasitic capacitance, C_p, of inductors is devised, it also improves the performance of voltage-controlled oscillators (VCOs). In the topology of Fig. 3.2, for example, reduction of C_p directly translates to a wider tuning range because the varactor diodes can contribute more variable capacitance. Simulations indicate that

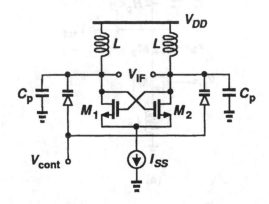

Figure 3.2.   Representative VCO.

the inductor modification introduced in this chapter increases the tuning range of a 900-MHz CMOS VCO from 4.2% to 23% for a 2x varactor capacitance range.

A candidate for compact, high-value inductors is the stacked structure of Fig. 3.3, originally introduced in GaAs technology [10] and later used in CMOS technology as well [11]. From the circuit model of Fig. 3.3, it

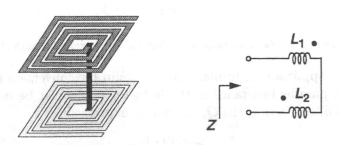

Figure 3.3.   A two-layer inductor.

can be seen that the input impedance of this structure is

$$Z = j\omega(L_1 + L_2 + 2M), \tag{3.5}$$

where L_1 and L_2 are the self-inductance of the spirals and M is the mutual inductance between the two. In a stacked inductor, the two spirals are identical ($L_1 = L_2 = L$) and the mutual coupling between the two layers is quite strong ($M \approx \sqrt{L_1 L_2} = L$). The total inductance is therefore increased by nearly a factor of 4. Similarly, for an n-layer inductor the total inductance is nominally equal to n^2 times that of one spiral. With the availability of more than five metal layers in modern CMOS technologies, stacking can provide increasingly larger values in a small area.

4. Derivation of Self-Resonance Frequency

Stacked structures typically exhibit a single resonance frequency. So, they can be modeled by a lumped RLC tank with $f_{SR} = (2\pi\sqrt{L_{eq}C_{eq}})^{-1}$, where L_{eq} and C_{eq} are the equivalent inductance and capacitance of the structure, respectively. While the equivalent inductance can be obtained by various empirical expressions [12, 13], Greenhouse's method [14], or electromagnetic field solvers [15], no method has been proposed to calculate the equivalent capacitance. We derive an expression for the capacitance in this section.

For f_{SR} calculations, we decompose each spiral into equal sections as shown in Fig. 3.4(a) such that all sections have the same inductance and parasitic capacitance to the substrate or the other spiral. This decomposition yields the distributed model illustrated in Fig. 3.4(b). In this circuit, inductive elements $L_{i,j}$'s represent the inductance of each section in Fig. (a) and they are all mutually coupled. The capacitance between the two layers is modeled by capacitors $C_{1,j}$ and that between the bottom layer and the substrate by capacitors $C_{2,j}$. To include the finite Q of the structure, all sources of loss are lumped into parallel resistor elements $R_{i,j}$. Also, we neglect trace-to-trace capacitances of each spiral. The validity of these assumptions will be explained later.

The simple circuit model of Fig. 3.4(b) still does not easily lend itself to current and voltage equations. However, we can use the physical definition of resonance. The resonance frequency can be viewed as the frequency at which the peak magnetic and electric energies are equal. In other words, if we calculate the total electric energy stored in the structure for a given peak voltage V_0 and equate that to $C_{eq}V_0^2/2$, then we can obtain C_{eq}.

To derive the electric energy stored in the capacitors, we first compute the voltage profile across the uniformly distributed capacitance of the

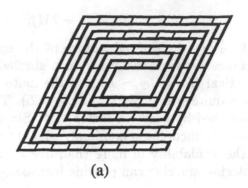

(a)

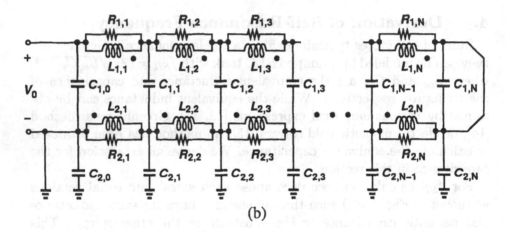

(b)

Figure 3.4. (a) Decomposing a spiral into equal sections. (b) Distributed model of a two-layer inductor.

structure. Assuming perfect coupling between every two inductors in Fig. 3.4, we express the voltage across each as:

$$V_{L,l,m} = \sum_{k=1}^{2} \sum_{n=1}^{N} j\omega I_{k,n} L_{k,n}, \qquad (3.6)$$

where $I_{l,m}$ is the current through $L_{l,m}$ and N is the number of the sections in the distributed model. Equation (3.6) reveals that all inductors sustain *equal* voltages. Therefore, for a given applied voltage V_0, we have

$$V_{L,l,m} = \frac{V_0}{2N}. \qquad (3.7)$$

From (3.6) and (3.7), it follows that the voltage varies linearly from V_0 to 0 across the distributed capacitance C_1 and from 0 to $V_0/2$ across C_2

(from left to right in Fig. 3.5). Having determined the voltage variation,

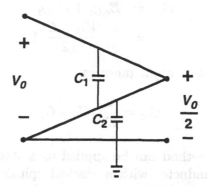

Figure 3.5. Voltage profile across each capacitor.

we write the electric energy stored in the mth element, $C_{1,m}$, as

$$E_{e,C_{1,m}} = \frac{1}{2}C_{1,m}[(V_0 - mV_{L,l,m}) - mV_{L,l,m}]^2. \tag{3.8}$$

The total electric energy in C_1 is therefore equal to

$$E_{e,C_1} = \frac{1}{2}\sum_{m=0}^{N} C_{1,m}(V_0 - 2mV_{L,l,m})^2. \tag{3.9}$$

As mentioned earlier, all sections are identical, i.e., $C_{1,m} = C_1/(N+1)$, and if we substitute (3.7) in (3.9), define a new variable $x = m/N$, and let N go to infinity, then we obtain

$$E_{e,C_1} = \frac{1}{2}C_1V_0^2\int_0^1 (1-x)^2 dx \tag{3.10}$$

$$= \frac{1}{2}\cdot\frac{C_1}{3}V_0^2. \tag{3.11}$$

The above equation states that if the voltage across a distributed capacitor changes linearly from zero to a maximum value V_0, then the equivalent capacitance is $1/3$ of the total capacitance. Since C_2 sustains a maximum voltage of $V_0/2$, its electric energy is equal to

$$E_{e,C_2} = \frac{1}{2}\cdot\frac{C_2}{3}\cdot(\frac{V_0}{2})^2 \tag{3.12}$$

$$= \frac{1}{2}\cdot\frac{C_2}{12}V_0^2. \tag{3.13}$$

From (3.11) and (3.13), the total electric energy stored in the inductor is

$$E_e = E_{e,C_1} + E_{e,C_2} \tag{3.14}$$

$$= \frac{1}{2} \cdot \frac{4C_1 + C_2}{12} V_0^2, \tag{3.15}$$

yielding the equivalent capacitance as

$$C_{eq} = \frac{1}{12}(4C_1 + C_2). \tag{3.16}$$

The foregoing method can be applied to a stack of multiple spirals as well. For an inductor with n stacked spirals, (3.6) suggests that the voltage is equally divided among the spirals. Therefore, interlayer capacitances sustain a maximum voltage of $2V_0/n$ whereas the bottom-layer capacitance sustains V_0/n. Now, using the result of (3.11) and adding the electric energy of all layers, we have

$$E_e = \frac{1}{2} \sum_{i=1}^{n-1} \frac{C_i}{3} (\frac{2V_0}{n})^2 + \frac{1}{2} \cdot \frac{C_n}{3} (\frac{V_0}{n})^2 \tag{3.17}$$

$$= \frac{1}{2} \cdot \frac{4 \sum_{i=1}^{n-1} C_i + C_n}{3n^2} V_0^2, \tag{3.18}$$

and hence

$$C_{eq} = \frac{1}{3n^2} (4 \sum_{i=1}^{n-1} C_i + C_n). \tag{3.19}$$

The simplified model used to derive the equivalent capacitance is slightly different from the exact physical model of a stacked inductor. The following three issues must be considered:

(1) We have assumed that all inductors in the distributed model are perfectly coupled. However, the coupling between orthogonal segments of a spiral or different spirals is very small. Nonetheless, if we assume that the inductor elements that are on top of each other are strongly coupled, then they sustain equal voltages. Therefore, the total voltage is still equally divided among the spirals. Furthermore, since each spiral is composed of a few groups of coupled inductors, the linear voltage profile is a reasonable approximation. To verify the last statement, a two-turn single spiral has been simulated. The spiral has been divided into 20 sections (12 sections for the outer turn and 8 sections for the inner turn). Then, inductor elements in the same segment and parallel

adjacent segments are strongly coupled while there is no magnetic coupling between other segments (orthogonal and parallel segments with opposite current direction). Figure 3.6 shows the voltage profile for this structure. As seen in this figure, the actual profile is relatively close to the linear approximation.

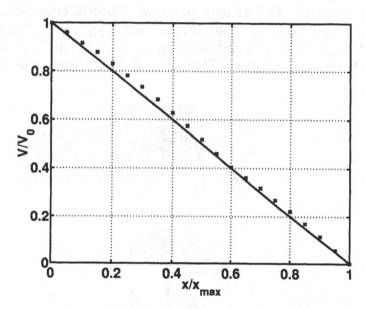

Figure 3.6. Simulated voltage profile of a single spiral.

(2) We have neglected the electric energy stored in the trace-to-trace capacitance, C_{TT} (the capacitance between two adjacent turns in the same layer). Supported by the experimental results in Section VII, this assumption can be justified by two observations. First, the width of the metal segments is typically much greater than the metal thickness. Therefore, even for a small spacing between the segments, C_{TT} is usually smaller than the interlayer capacitance. Second, the adjacent turns in the same spiral sustain a small voltage difference. Noting that the electric energy is proportional to the square of voltage, we conclude that the effect of C_{TT} is negligible.

(3) Presenting all of the loss mechanisms by parallel resistors in the distributed model introduces little error in the calculation of the self-resonance frequency. For metal resistance and magnetic coupling to the substrate, parallel resistors are a good model if $Q^2 \gg 1$.

It is important to note that measurements indicate that Eq. (3.19) provides a reasonable approximation for f_{SR} of a *single* spiral as well, though the focus of the chapter is on stacked spirals.

5.   Modification of Stacked Inductors

For a two-layer inductor, (3.16) reveals that the interlayer capacitance, C_1, impacts the resonance frequency four times as much as the bottom-layer capacitance, C_2. In addition, for two adjacent metal layers, C_1 is several times greater than C_2. Therefore, it is plausible to move the spirals farther from each other so as to achieve a higher self-resonance frequency. For example, in a typical CMOS technology with five metal layers, $C_{M_5-M_4} \cong 40$ aF/μm^2 and $C_{M_4-sub} \cong 6$ aF/μm^2, whereas $C_{M_5-M_3} \cong 14$ aF/μm^2 and $C_{M_3-sub} \cong 9$ aF/μm^2. It follows that for the structure of Fig. 3.7(a), $C_{eq,a} \approx 14$ aF/μm^2, whereas for

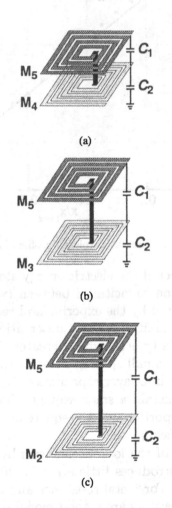

Figure 3.7.   Modification of two-layer stacked inductors.

Fig. 3.7(b), $C_{eq,b} \approx 5.4$ aF/μm^2, an almost three-fold reduction.

Equation (3.16) proves very useful in estimating the performance of various stack combinations. For example, it predicts that the structure of Fig. 3.7(c) has an equivalent capacitance $C_{eq,c} \approx 4$ aF/μm^2 because $C_{M5-M2} \cong 9$ aF/μm^2 and $C_{M2-sub} \cong 12$ aF/μm^2. In other words, the self-resonance frequency of the inductor in Fig. 3.7(c) is almost twice that of the inductor in Fig. 3.7(a).

Note that the value of the inductance remains relatively constant because the lateral dimensions are nearly two orders of magnitude greater than the vertical dimensions. By the same token, the loss through the substrate remains unchanged. Both of these claims are confirmed by measurements (Section VII).

The idea of moving stacked spirals away from each other so as to increase f_{SR} can be applied to multiple layers as well. For example the structure of Fig. 3.8(a) can be modified as depicted in Fig. 3.8(b), thereby raising f_{SR} by 50%.

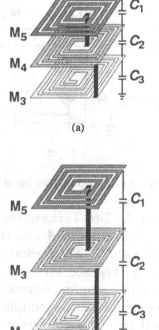

Figure 3.8.   Three-layer stacked inductor modification.

6.   Stacked Transformers

Monolithic transformers producing voltage or current gain can serve as interstage elements if the signals do not travel off chip, i.e., if power gain is not important. Such transformers can also perform single-ended to differential and differential to single-ended conversion.

A particularly useful example is depicted in Fig. 3.9, where a transformer having current gain is placed in the current path of an active mixer. Here, the RF current produced by M_1 is amplified by T_1 before it is commutated to the output by M_2 and M_3. The current gain lowers the noise contributed by M_2 and M_3 and it is obtained with no power, linearity, or voltage headroom penalty.

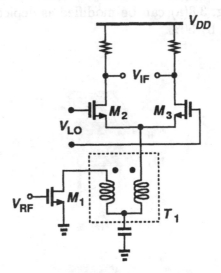

Figure 3.9.   Example of using a transformer to boost current in an active mixer.

Figure 3.10(a) shows the 1-to-2 transformer structure. The primary is formed as a single spiral in metal 4 and the secondary as two series spirals in metal 3 and metal 5. The performance of the transformer is determined by the inductance and series resistance of each spiral and the magnetic and capacitive coupling between the primary and the secondary. To minimize the capacitive coupling, the primary turns are offset with respect to the secondary turns as illustrated in Fig. 3.10(b). Thus, the capacitance arises only from the fringe electric field lines. The number of turns in each spiral also impacts the voltage (or current) gain at a desired frequency because it entails a trade-off between the series resistance and the amount of magnetic flux enclosed by the primary and the secondary. For single-ended to differential conversion, two of the

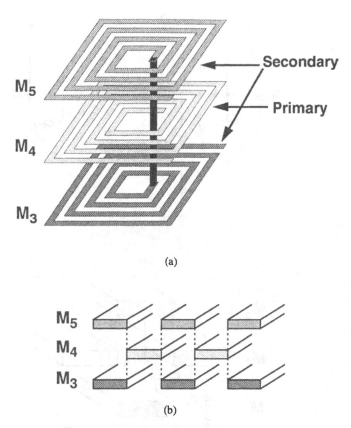

(a)

(b)

Figure 3.10. Transformer structure.

structures in Fig. 3.10(a) can be cross-coupled so as to achieve symmetry.

To design the transformer for specific requirements, a circuit model is necessary. Figure 3.11 illustrates one section of the distributed model developed for the 1-to-2 transformer. The segments L_i and R_i represent a finite element of each spiral, C_f's denote the fringe capacitances, C_1 models the capacitance between M_5 and M_3, and C_2 and C_3 are the capacitances between the substrate and M_3 and M_4, respectively. The values of L_i and R_i are derived assuming a uniformly-distributed model and a Q of 3 for each inductor. The capacitance values are obtained from the foundry interconnect data. Figure 3.12 depicts the simulated voltage gain of two transformers, one consisting of eight-turn spirals with 7-μm wide metal lines and the others consisting of four-turn and three-turn spirals with 9-μm wide metal lines.

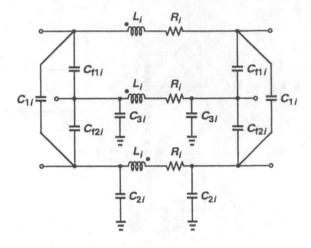

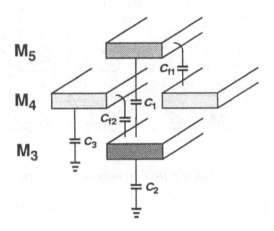

Figure 3.11.   Transformer model.

Unlike stacked inductors, whose resonance frequency is not affected by the inductor loss, the transfer characteristics and voltage gain of the transformer depend on the quality factor of the spirals. In this simulation, a Q of 3 has been used for each winding. As Fig. 3.12 shows, for the eight-turn transformer, capacitive coupling between the spirals is so large that it does not allow the voltage gain to exceed one, while for the four-turn and three-turn transformers we expect a gain of about 1.8 in the vicinity of 2 GHz. Note that if the secondary is driven by a current source and the short-circuit current of the primary is measured, the same characteristics are observed.

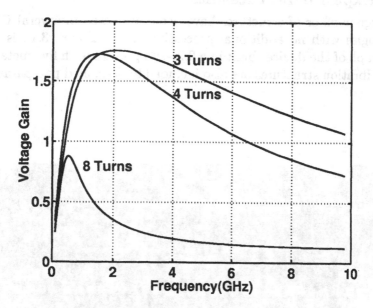

Figure 3.12. Simulated voltage gain of the transformers.

The concept of stacked transformer can be applied to more layers of metal to achieve higher voltage gains. Figure 3.13 shows a stacked transformer with a nominal gain of 4. In this structure, M_3 forms the primary and the rest of the metal layers are used for the secondary.

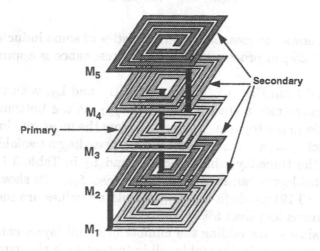

Figure 3.13. 1-to-4 transformer structure.

7.   Experimental Results

A large number of structures have been fabricated in several CMOS technologies with no additional processing steps. Figure 3.14 is a die photograph of the devices built in a 0.25-μm process with five metal layers. Calibration structures are also included to de-embed pad parasitics.

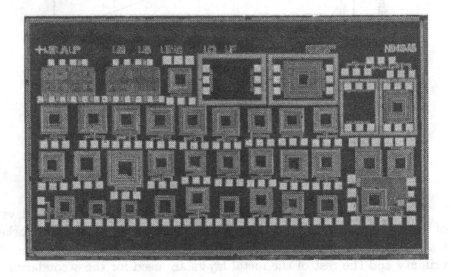

Figure 3.14.   Die photo.

Table 3.1 shows the measured characteristics of some inductors fabricated in the 0.25-μm process. The Q at self-resonance is approximately equal to 3.

As expected from Fig. 3.7, inductors L_1, L_2, and L_3, with two layers of metal, demonstrate a steady increase in f_{SR} as the bottom spiral is moved away from the top one. Figure 3.15 plots the measured impedance of these inductors as a function of frequency, revealing a twofold increase in f_{SR}. For the three-layer inductors (L_4 and L_5 in Table 3.1), proper choice of metal layers can considerably increase f_{SR}. To show how accurately Eq. (3.19) predicts the f_{SR}, calculated values are included as well. The error is less than 5%.

Table 3.2 shows how adding the number of metal layers can increase the inductance value. In this table, all inductors have the same dimensions but incorporate different number of layers. Using five layers of metal yields an inductance value of 266 nH in an area of $(240\ \mu m)^2$.

Inductor	Metal Layers	L (nH)	Measured f_{SR} (GHz)	Calculated f_{SR} (GHz)	Number of Turns
$L_1(240\mu m)^2$	5,4	45	0.92	0.96	7
$L_2(240\mu m)^2$	5,3	45	1.5	1.53	7
$L_3(240\mu m)^2$	5,2	45	1.8	1.79	7
$L_4(240\mu m)^2$	5,4,3	100	0.7	0.7	7
$L_5(240\mu m)^2$	5,3,1	100	1.0	1.0	7
$L_6(200\mu m)^2$	5,3,2	50	1.5	1.46	5
$L_7(200\mu m)^2$	5,2,1	48	1.5	1.54	5

Table 3.1. Measured inductors in 0.25-μm technology (linewidth= 9 μm, line spacing= 0.72 μm).

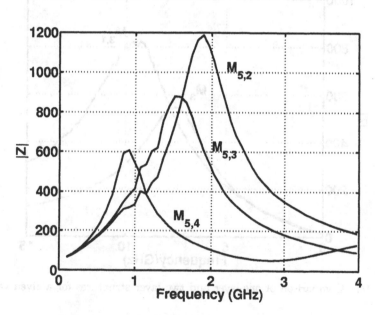

Figure 3.15. Measured inductor characteristics.

Accommodating such high values in a small area makes these inductors attractive for integrating voltage regulators and dc-dc converters monolithically.

Stacking inductors can also be useful even for small values. Figure 3.16 shows two 5-nH inductors fabricated in a 0.6-μm technology with three layers of metal. The two inductors were designed for the same inductance and nearly equal Q's. The plots in Fig. 3.16(b) show that the stacked structure has a higher f_{SR} because it occupies less area.

Inductor Size	Metal Layers	L (nH)	Measured f_{SR}(GHz)	Calculated f_{SR}(GHz)
$(240\mu m)^2$	5,4	45	0.92	0.96
$(240\mu m)^2$	5,4,3	100	0.7	0.7
$(240\mu m)^2$	5,4,3,2	180	0.55	0.58
$(240\mu m)^2$	5,4,3,2,1	266	0.47	0.49

Table 3.2. High-value inductors in 0.25-μm technology (linewidth= 9 μm, line spacing= 0.72 μm, number of turns for each spiral= 7).

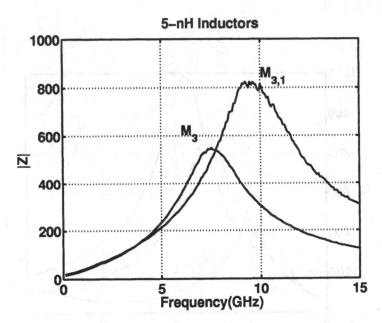

Figure 3.16. Comparison of one-layer and two-layer structures for a given value of inductance.

In Fig. 3.17, some other measured results for two pairs of 5-nH and 10-nH inductors in a 0.4-μm technology (with four layers of metal) are presented. In this case, the self-resonance frequency increases by 50% with the proposed modification. The Q at self-resonance is between 3 and 5 for the four cases. Note that for the 5-nH inductor resonating at 11.2 GHz, the skin effect is quite significant. Measured and calculated values of f_{SR} [from (3.19)] differ by less than 4%.

As mentioned before, with the proposed modification, the inductance remains relatively constant because the lateral dimensions are nearly

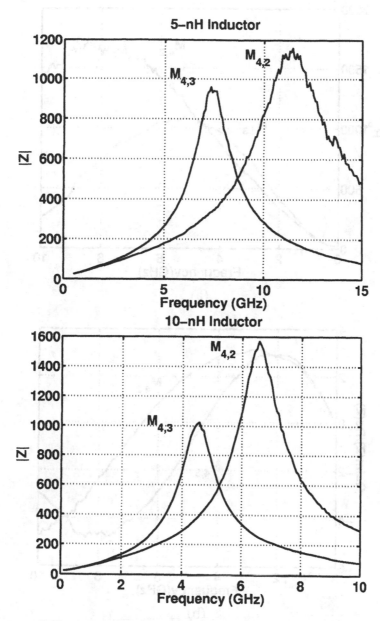

Figure 3.17. Measured inductors in 0.4-μm technology.

two orders of magnitude greater than the vertical dimensions. This is indeed evident from the slope of $|Z|$ at low frequencies, which is equal to $2\pi L$ (Figs. 3.15 and 3.17).

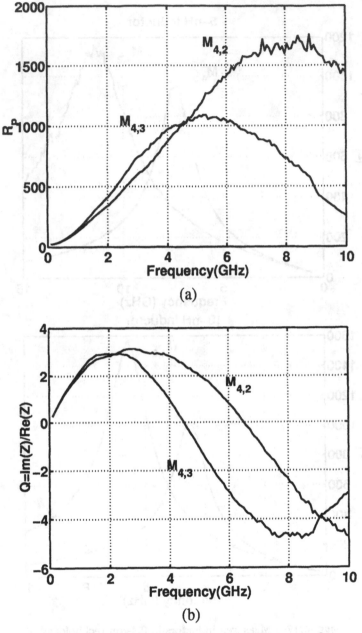

Figure 3.18. Effect of inductor modification on Q.

The effect of the proposed modification on the Q is also studied. For the two 10-nH inductors of Fig. 3.17, we can derive the parallel resistance, R_p, as a function of frequency [Fig. 3.18(a)]. If the Q is defined as in (3.3), then the two inductors have equal Q's around 5 GHz and if Eq. (3.2) is used, the Q's are even closer for frequencies below the resonance [Fig. 3.18(b)].

Perhaps a fairer comparison is to assume each of the inductors is used in a circuit tuned to a given frequency (e.g., as in a VCO). We then add enough capacitance to the modified structure so that it resonates at the same frequency as the conventional one. Figure 3.19 shows that the two inductors have the same selectivity and hence the same Q, while the modified structure can sustain an additional capacitance of 87 fF for operation at 4.5 GHz.

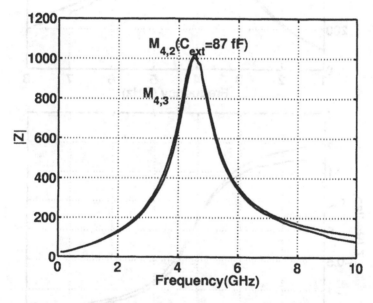

Figure 3.19. Effect of inductor modification on selectivity.

To simulate the behavior of an inductor, we can use the distributed circuit of Fig. 3.3 with a finite number of sections (e.g., 10). However, measured results indicate that for tuned applications, stacked inductors can be even modeled by a simple parallel RLC tank. Figure 3.20 compares the simulation results of a parallel RLC tank and the measured characteristics.

Here, the equivalent capacitance obtained from (3.16) and the measured value of the parallel resistance at the resonance frequency are used. These plots suggest that the magnitudes are nearly equal for a wide range and the phases are close for about ±10% around resonance.

Several 1-to-2 transformers have been fabricated in a 0.25-μm technology. Figure 3.21 plots the measured voltage gains as a function of

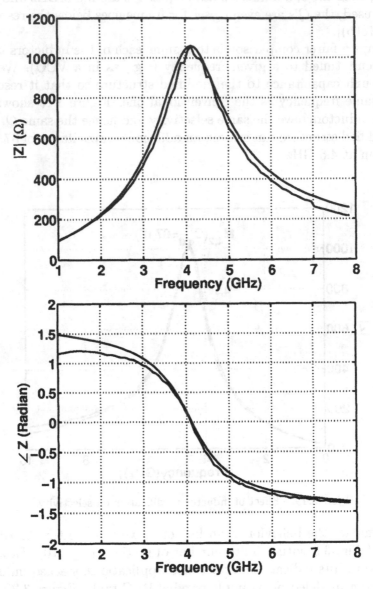

Figure 3.20. Simulation and measurement comparison.

frequency. The measured behavior is reasonably close to the simulation results using the distributed model. The four-turn transformer achieves a voltage gain of 1.8 at 2.4 GHz and the three-turn transformer has nearly the same voltage gain over a wider frequency range. The plot also illustrates the effect of capacitive loading on the secondary (calcu-

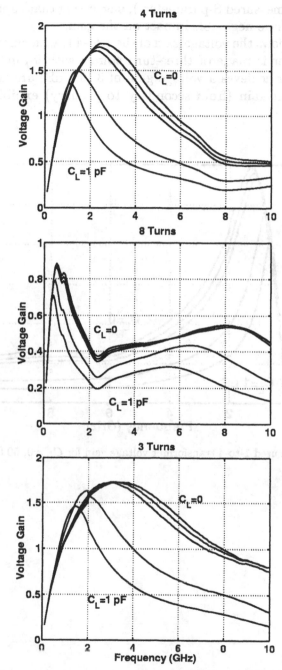

Figure 3.21. Measured 1-to-2 transformer voltage gain for $C_L = 0$, 50 fF, 100 fF, 500 fF, 1 pF.

lated using the measured S-parameters), suggesting that capacitances as high as 100 fF have negligible impact on the gain.

Figure 3.22 shows the voltage gain of the 1-to-4 transformer of Fig. 3.13. This transformer is made of three-turn spirals with 9-μm metal lines. The transformer achieves a voltage gain of 3 (9.5 dB) around 1.5 GHz. The short-circuit gain (from secondary to primary) exhibits identical characteristics.

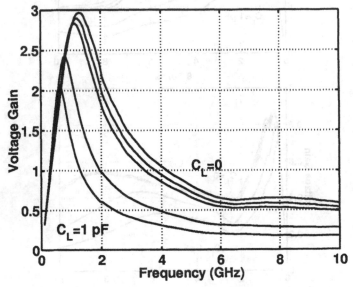

Figure 3.22.   Measured 1-to-4 transformer voltage gain for C_L =0, 50 fF, 100 fF, 500 fF, 1 pF.

Chapter 4

RECEIVER FRONT END

1. Introduction

The RF front end is a critical building block in the receiver that greatly affects the sensitivity and linearity of the system. The receiver front end described here consists of a low-noise amplifier (LNA) and two downconversion mixers. In this chapter we start with a brief study of LNAs and mixers in CMOS technology. We then describe stacking techniques to reduce power consumption and present downconversion stages.

2. Low-Noise Amplifier

There are two types of methods commonly used to design an LNA for RF CMOS circuits: common-gate and cascode amplifiers. While the common-gate stage provides a wide-band input matching and is less sensitive to parasitics [23], it has an inherently high noise figure. If the channel noise of a MOS transistor is specified as $4kT\gamma g_m$, where γ is the excess noise factor and g_m is the transconductance of the device, the noise factor of the amplifier will be [1]

$$F = 1 + \gamma. \qquad (4.1)$$

Although γ is 2/3 in long channel devices, it can be quite high in short channel transistors (e.g., 2.5), leading to a high noise figure. Therefore, in most applications where the noise factor is a critical issue, a cascode LNA with inductive degeneration is preferable. Figure 4.1 shows such a circuit. Neglecting the gate-drain capacitance and the output resistance

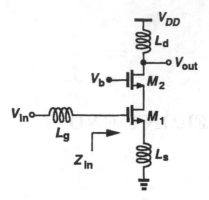

Figure 4.1. Cascode LNA.

of the transistor, we can express the input impedance of the LNA as

$$Z_{in}(s) = L_s \frac{g_m}{C_{gs}} + \frac{1}{C_{gs}s} + L_s s \qquad (4.2)$$

Thus, the input resistance of the amplifier (the real part of the input impedance) is

$$R_{in} = L_s \frac{g_m}{C_{gs}} \qquad (4.3)$$

$$= L_s \omega_T \qquad (4.4)$$

where ω_T is the unity gain frequency of the device. By proper choice of g_m, L_s and C_{gs}, the input resistance can be equal to the 50-Ω source resistance and the input reactance (imaginary part of impedance) can be resonated out by a series inductor (L_g).

In this circuit, if the input impedance is matched to the source resistance, i.e., $R_s = g_m L_s/C_{gs}$ and $L_g \omega = 1/C_{gs}\omega - L_s\omega$, then, the noise factor can be specified as

$$F = 1 + \gamma \frac{R_s C_{gs}^2 \omega^2}{g_m} \qquad (4.5)$$

$$= 1 + \gamma g_m R_s (\frac{\omega}{\omega_T})^2. \qquad (4.6)$$

This expression shows that for a given ω_T, as we lower the bias current, the noise factor decreases! This is in fact true because if we define the quality factor of the input matching circuit as

$$Q = \frac{1}{2R_s C_{gs}\omega}, \qquad (4.7)$$

(when the circuit is matched, the total series resistance is $2R_s$) then, the effective transconductance of the circuit, G_m, is [23]

$$G_m = g_m \cdot Q \tag{4.8}$$

$$= \frac{g_m}{2R_sC_{gs}\omega} \tag{4.9}$$

$$= \frac{1}{2R_s}\frac{\omega_T}{\omega}. \tag{4.10}$$

Therefore, for a given ω_T, the effective transconductance of the circuit is constant, while as we lower the bias current, the output noise decreases. In other words, for a given ω_T, a lower bias current leads to a lower noise factor. Even though the above statement makes common-source amplifiers very attractive for low-power design, practical problems limit the minimum bias current of the device. As we lower the bias current, while keeping ω_T constant, C_{gs} decreases, leading to a higher Q. A high-Q matching network has several drawbacks. First, the circuit becomes very sensitive to component variations and parasitics. Second, the off-chip matching circuit which in this case contains a series inductor, inserts a large amount of loss at the input (even for a high-Q off-chip inductor). For example, in an extreme case where the bias current and C_{gs} approach zero, theoretically we expect a noise figure of 0 dB while the insertion gain is determined by ω_T/ω. However, the series inductor L_g needed to match the input, goes to infinity, giving infinite loss at the input of the LNA. Finally, using a high Q matching network degrades the linearity of the LNA, which is determined by the voltage gain from the input source to the gate-source voltage and the overdrive voltage of the device ($V_{gs} - V_{TH}$). In practice, a Q of about 3 is acceptable and (4.6) can be expressed as

$$F = 1 + \frac{\gamma}{4Q^2} \cdot \frac{1}{g_mR_s}. \tag{4.11}$$

Therefore, for a given Q, higher g_m improves the noise figure.

The noise factor of a cascode stage can also be specified in terms of C_{gs} and L_s. Replacing the source impedance from (4.3) into (4.5) yields

$$F = 1 + \gamma C_{gs}L_s\omega^2 \tag{4.12}$$

$$= 1 + \gamma(\frac{\omega}{\omega_s})^2 \tag{4.13}$$

where ω_s is the resonance frequency of C_{gs} and L_s, which is quite higher than the frequency of operation, ω. Comparing (4.13) and (4.1), we can see how a common source stage improves the noise figure.

Thus far, we have assumed that the source resistance is equal to the input resistance seen at the gate of the transistor given by (4.4). For a given source resistance of 50 Ω, as we reduce L_s, ω_T increases but the minimum value of L_s is limited by parasitic and sensitivity issues. For example, suppose that L_s is made of two parallel wirebond inductors with an equivalent inductance of 1 nH. In order to provide a 50-Ω input resistance, ω_T needs to be $2\pi \times 8$ GHz, which is far below the technology limit and degrades the noise figure. Even increasing the bias current does not help because as (4.6) shows, for a given Ω_T set by (4.4), increasing the current results in a higher g_m, giving a higher noise figure.

To resolve this problem, we can use a matching circuit that boosts the impedance seen at gate of the transistor looking into the input source (Fig. 4.2(a)). Figure 4.2(b) shows an example of an L-match network for this purpose [33]. When a lossless passive circuit provides power

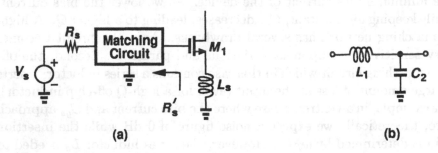

Figure 4.2. Matching circuit to boost the impedance seen at the gate of the transistor looking into the source.

matching at the source node, all nodes in the input path will be power matched. Therefore, all equations that were derived are valid if R_S is replaced by R'_s and Q is replaced by $Q' = 1/(2R'_s C_{gs})$.

It must be noted that the actual Q of the circuit in this case is that seen at the input of the device (Q') times the voltage gain of the matching circuit, which is $\sqrt{R'_s/R_s}$. Thus, in order to keep the Q constant, Q' needs to be reduced by a factor of $\sqrt{R'_s/R_s}$, decreasing C_{gs} by this factor. Now that we have achieved to reduce C_{gs} for a given Q and L_s, the noise factor decreases as shown by (4.13). This is in fact at the cost of the extra power consumption for a $\sqrt{R'_s/R_s}$-fold increase in g_m. If we assume a square-law I-V equation for the input device, and a constant (minimum) channel length, the bias current has to be scaled up by $(R'_s/R_s)^{3/2}$, while the overdrive voltage increases by a factor of (R'_s/R_s), providing more linearity.

In this analysis, we have only considered the simplified effect of the channel noise in the input device. Although this is the dominant factor determining the noise of the LNA, in practice, there are other factors that increase the noise. These include the gate-induced noise [20], effect of gate-drain capacitance [21] and gate resistance noise [22].

Another important source of noise in the cascode topology is the noise introduced by the cascode device, M_2, added to improve the stability of the amplifier. At low frequencies, if channel length modulation in M_1 is neglected, M_2 does not affect the noise of the amplifier. At high frequencies, on the other hand, the capacitance at the drain of M_1, reduces the impedance at this node, increasing the output noise. Since the noise contribution of the M_2 is proportional to the capacitance at this node, it is very important to minimize this capacitance. One way to minimize this capacitance is merging the drain of M_1 into the source of M_2 in layout. This allows eliminating the metal contacts at this node, reducing the junction area. However, this is only possible when the widths of both devices are the same. Nevertheless, simulation results show that using the same width and merging the junctions can be more effective than optimizing M_2 independently.

3.   RF Mixers

CMOS mixers can be implemented in two different forms: active and passive topologies. While passive mixers are more linear and do not need any dc power, there are a few issues that limit their usage for the first downconversion. First, they insert loss in the signal path, degrading the noise performance of the receiver. Second, they require large LO swings. Third, passive mixers perform better when they drive high impedance loads, such as small capacitances at low frequencies. On the other hand, the first IF in the proposed transceiver is 800 MHz, necessitating active mixers for the first downconversion.

Figure 4.3 shows a single-balanced active mixer. Since the IF is quite high, inductive loads are used. If we assume abrupt switching in M_2 and M_3, the gain of the mixer can be approximated by

$$A_v = \frac{2}{\pi} g_{m1} R_P, \qquad (4.14)$$

where R_P is the equivalent parallel resistance of the inductive loads. The noise of the mixer can also be estimated by [7]

$$\frac{v_o^2}{\Delta f} = 8kTR_P \left(1 + \gamma \frac{R_P I}{\pi A} + \frac{g_m R_P}{2}\right) \qquad (4.15)$$

where I is the bias current of the input device and A is the amplitude of the LO signal. As (4.15) shows, when the bias current I is increased, the noise of the transconductor device is reduced while the noise of the switches increases. Another interesting result from (4.15) is that the

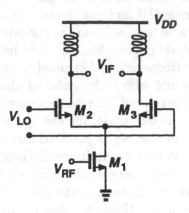

Figure 4.3. Single-balanced active mixer.

noise of the switches only depends on the bias current and the LO amplitude. Therefore, the size of the switches does not affect the noise performance. In practice, however, the dimensions of the switches affect both the conversion gain and the parasitic capacitance at the drain of M_1.

4.   First Downconversion

Figure 4.4 shows a conventional approach to the first downconversion. The circuit consists of a cascode LNA capacitively coupled to a single-balanced active mixer. While providing a simple solution for the first downconversion, this circuit suffers from three important drawbacks. First, since there are two current paths from the supply, the power consumption is high. Second, the image rejection of the circuit is relatively low. This is due to the low quality factor of on-chip inductors in CMOS technology. Simulation results indicate that the image rejection given by the tank at the output of the LNA is limited to 20 dB. Third, the mixer is single-balanced, hence, the LO signal is mixed with the dc current of the mixer input device, creating LO feedthrough at 1.6 GHz at the IF port. Since the IF port is tuned at 800 MHz, the on-chip inductors at the output of the mixer cannot sufficiently suppress the LO feedthrough. As a result, the following stages can be desensitized.

To reduce the power consumption, we can lower the supply voltage but the minimum supply voltage is dictated by headroom and linearity issues in baseband sections, the prescaler of the synthesizer, and the tuning range of the VCO. Thus, we set the supply voltage at 2.5 V and employ stacking techniques to lower the power consumption.

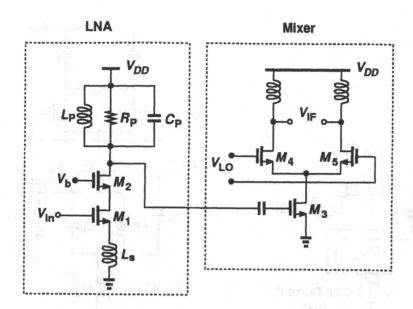

Figure 4.4. Conventional circuit for the first downconversion.

In the first downconversion, we can reuse the current of the LNA in the mixer by stacking the mixer on top of the LNA. This is shown in Fig. 4.5(a). Here, the signal travels through the path shown in black, while the supply current flows through the path shown in gray. The implementation is shown in Fig. 4.5(b). In this circuit, the LNA is stacked under the mixer and bypass capacitor C_b provides ac ground at the source of M_3. The RF signal at the output of the LNA is capacitively coupled to the input of the mixer through C_C. It must be noted that stacking is possible because the IF is high, allowing inductive loads rather than resistive loads and eliminating the headroom voltage of the loads.

To improve the image rejection of the circuit, we can modify the mixer as shown in Fig. 4.6. In this circuit, inductor L_i resonates at the image frequency, thus reducing the transconductance of M_3 for the image signal and providing another 20 dB of image rejection. This tank also suppresses the image noise of the LNA and mixer, improving the noise figure of the mixer by about 3 dB.

Finally, to resolve the LO feedthrough issue, the mixer can be modified to a double-balanced topology as shown in Fig. 4.7. In this circuit the input of the mixer is not fully differential. Nevertheless, the LO feedthrough is substantially reduced.

This stacking technique provides high performance for low power applications. For the device dimensions given in Table 4.1, simulation

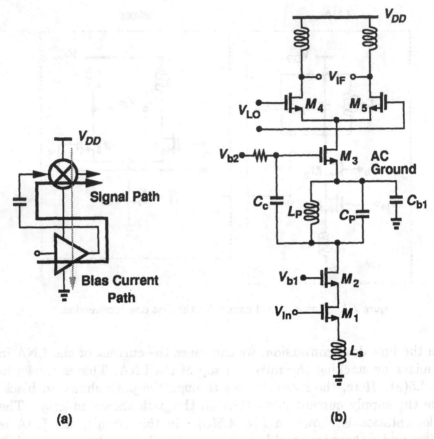

(a)   **(b)**

Figure 4.5.   (a) Stacking in the first downconversion (b) circuit implementation.

results show that with a supply current of 2.5 mA, the circuit achieves a noise figure of 3 dB and a total gain of 29 dB. In this simulation, all sources of noise are lumped into the channel noise with a γ of 2.5.

Device	$\frac{W}{L}$ ($\frac{\mu m}{\mu m}$)	g_m (mΩ^{-1}) (I_{supply}=2.5 mA)
M_1	100/0.25	17.4
M_2	100/0.25	18.5
M_3-M_4	30/0.25	6.4
M_5-M_8	50/0.25	6.8

Table 4.1.   Device dimensions in the first downconversion.

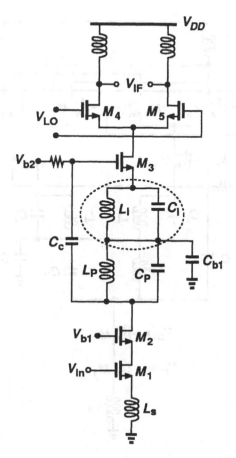

Figure 4.6. Improving image rejection in the first downconversion.

The circuit of Fig. 4.7 incorporates a number of passive components such as inductors and capacitors. Figure 4.8 shows the inductors used in the first downconversion circuit. The RF inductor at the output of the LNA is a two-layer 7-nH stacked spiral and the IF and image rejection inductors are four-layer 50-nH stacked spirals. In both inductors, a patterned n-well layer is placed under each structure. This layer acts as a shield for the electric field. Another source of loss in on-chip inductors is the eddy current in the substrate. In some processes, in order to increase the threshold voltage under the field oxide, a heavily-doped layer is used which can incur a large eddy loss. The patterned n-well layer also reduces this loss. The quality factor of inductors is about 3.5.

Two other passive components used in the first downconversion circuit are the coupling and bypass capacitors. The coupling capacitor used in

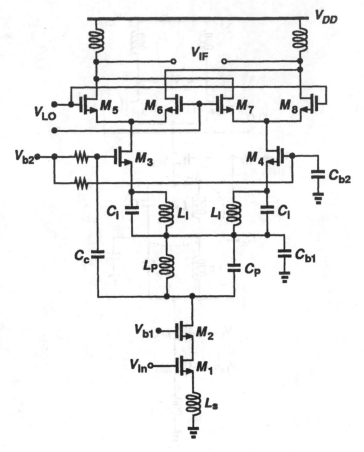

Figure 4.7. First downconversion circuit.

this circuit is made of a three-layer metal sandwich. This helps to reduce the area and the bottom-plate capacitance. Compared to a simple M_5-M_4 capacitor, the bottom-plate capacitance reduces from %17 to %12 in M_5-M_4-M_3 sandwich and the area is reduced to half. The connection of the coupling capacitor is such that the bottom-plate capacitance is resonated out by the inductor at the output of the LNA. Note that we use a standard digital CMOS process. So, metal-insulator-metal (MIM) capacitors are not available.

Another critical component in this circuit is the bypass capacitor. This capacitor must be large enough to avoid current coupling to the source of the mixer input device. To implement large capacitance values while occupying a reasonable chip area, MOS transistors with the source and drain connected to ground are used. Since this capacitor is

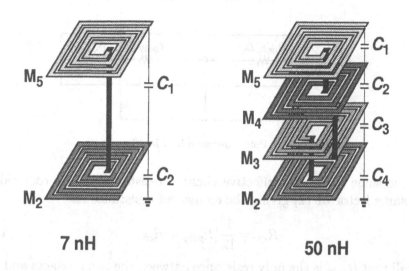

7 nH   **50 nH**

Figure 4.8. Inductors used in downconversion.

to provide ac ground at 2.4 GHz, the dimensions of the device need to be carefully designed to ensure a high quality factor at RF. The quality factor in this case is defined as

$$Q = \frac{1}{R_{eq}C\omega}. \tag{4.16}$$

where R_{eq} is the equivalent series resistance associated with capacitor C.

If the capacitor is composed of M fingers, each finger is designed so that its Q is large enough. After these fingers are connected together in an M-finger structure (parallel connection), the total capacitance is increased M times while the equivalent resistance is reduces by this factor. Therefore, the Q remains the same.

The equivalent series resistance, R_{eq}, is composed of two parts: poly resistance, R_{poly}, and channel resistance, R_{ch}. As shown in Fig. 4.9, for a poly strip with contacts on both sides, the effective poly resistance from contacts to the middle of the transistor is reduced by a factor of four (two $R_{poly}/2$'s in parallel). Since R_{poly} is a distributed resistance, the effective resistance from contacts to poly is reduced by another factor of 3, thus yielding an effective poly resistance of

$$R_{poly,eff} = \frac{1}{12}R_{poly}. \tag{4.17}$$

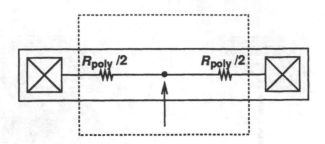

Figure 4.9. Poly resistance in a MOS capacitor.

By the same token, the effective channel resistance is also reduced by the same factor of 12, giving the equivalent resistance as

$$R_{eq} = \frac{1}{12}(R_{poly} + R_{ch}).$$
(4.18)

Recall that R_{poly} is the poly resistance between the two contacts and R_{ch} is the channel resistance between the source and drain. In this circuit, the bypass capacitor is designed to provide a Q of 20 at 2.4 GHz.

5.   Second Downconversion

For the second down conversion we must decide between active and passive mixers in terms of noise, linearity, power consumption and the required LO swings. While in the first downconversion, noise performance is more important, in the second downconversion, linearity is more critical. After careful analysis, we chose passive mixers because they are more linear and we were also able to design a low power divide-by-two circuit that drives the passive mixers. In addition, using passive mixers reduces power consumption and the flicker noise at the output.

5.1.   Divide-by-Two Circuit

Figure 4.10 shows the divide-by-two circuit. The circuit consist of two D-latches connected in a loop. To see how this circuit works, we can refer to Fig. 4.11, where waveforms are shown. As waveforms show, the frequency at the output is half of that at the input and the two outputs have a phase difference of $\pi/2$. This provides quadrature LO's and allows us to avoid power-hungry poly phase filters.

The D-latch circuit is depicted in Fig. 4.12. To maximize the output swings, cross-coupled PMOS devices are used in D-latches. Simulation results show that with a supply current of 1.5 mA, we can provide rail-to-rail swings to drive the passive mixers. Figure 4.13 shows the I and Q outputs of the divider when loaded by passive mixers.

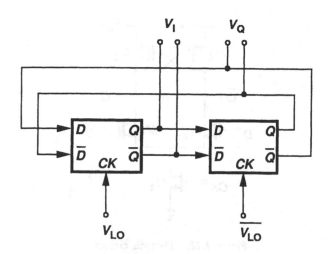

Figure 4.10. Divide-by-2 circuit.

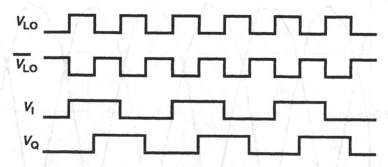

Figure 4.11. Waveforms in the divide-by-2 circuit.

5.2. Passive Mixers

Figure 4.14 shows the passive mixer used for the second downconversion. Bias voltages of the drain and source with respect to the gate voltage are important factors that affect the performance of the mixer. Since the output of the preceding stage is biased at V_{dd}, coupling capacitors are used to alleviate the biasing of the passive mixers. These capacitors also act as degenerative impedances in series with switches, improving the linearity of the mixer and preventing short circuits in the I and Q paths.

Since passive mixers cannot drive low-impedance resistive loads, they are followed by common source amplifiers as shown in Fig. 4.14. These amplifiers can limit the linearity of the system, but depending on the ap-

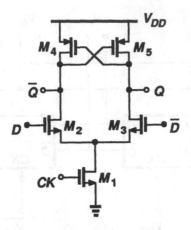

Figure 4.12.   D-latch circuit.

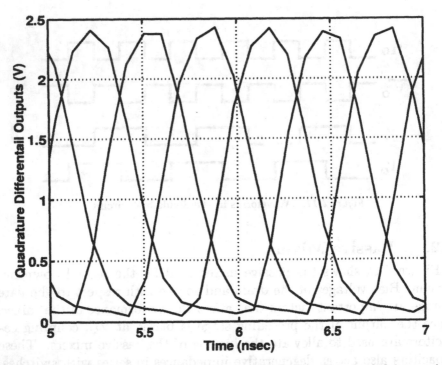

Figure 4.13.   Differential I and Q outputs of the divider when loaded by passive mixers.

plication, they can be eliminated and the passive mixers can be directly connected to variable-gain amplifiers (VGA) or channel-select filters.

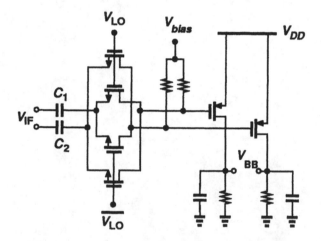

Figure 4.14. Second downconversion circuit.

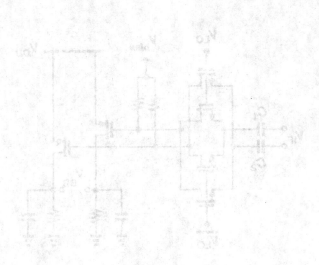

Chapter 5

TRANSMITTER

1. Introduction

Figure 5.1 shows the transmitter block diagram. As explained in

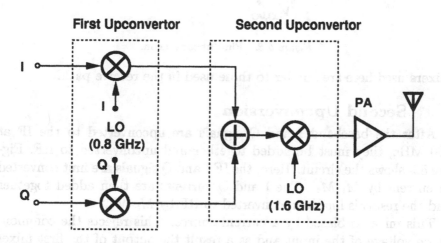

Figure 5.1. Transmitter block diagram.

Chapter 2, the I and Q baseband signals are first upconverted to 800 MHz and then added together. In the second mixer, the result is upconverted from 800 MHz to 2.4 GHz. After two upconversions, the RF signal is applied to a power amplifier (PA) to drive the antenna. In this chapter, we describe the design of the transmitter building blocks.

2.   First Upconversion

The first upconversion stage can significantly impact the distortion of the transmit path. This is due to the fact that the harmonics generated in this stage are close to the desired signal and will appear as sideband spurs at RF whereas the harmonics produced by the second mixer and the PA are quite out of band.

Since the distortion of this stage is critical, we have used passive mixers to upconvert the baseband I and Q signals. The circuit is shown in Fig. 5.2. Using passive mixers also helps to save power. The passive

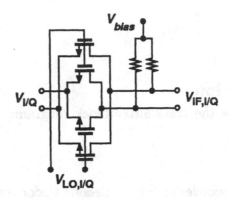

Figure 5.2.   First upconversion.

mixers used here are similar to those used in the receive path.

3.   Second Upconversion

After the baseband I and Q signals are upconverted to the IF at 800 MHz, they must be added together and upconverted to RF. Figure 5.3 shows the circuit. Here, the IF I and Q signals are first converted to current by M_1-M_4. The I and Q currents are then added together and the result is finally upconverted to RF by M_5-M_8.

This mixer is biased by a current source. This rejects the common-mode voltage of the input and as a result the output of the first mixer is directly connected to the second upconverter (with no capacitive coupling). To improve the linearity of the input transconductors, a bypass capacitor is used in parallel with the current source.

After upconversion to RF, the RF current must be converted to voltage again. The PA is typically a single-ended circuit, while the output of the mixer is differential. A common technique to convert a differential signal to a single-ended signal is using a tuned current mirror [25] as shown in Fig. 5.4. However, the main difficulty here is the large gate-

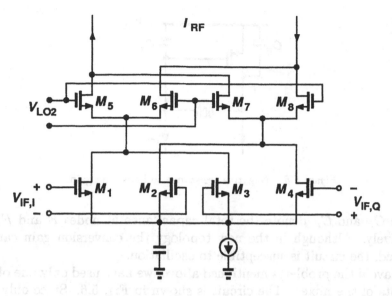

Figure 5.3. Second upconversion.

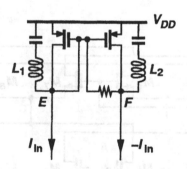

Figure 5.4. Current mirror to convert differential to single-ended signal.

source capacitance of the two PMOS devices, dictating a small value for L_1 that leads to a low conversion gain.

An alternative topology to increase the gain is using a negative resistance in parallel with a floating inductor as shown in Fig. 5.5 [25]. It can be shown that [1]

$$Z_{in}(s) = \frac{g_{m1}}{C_E C_F s^2} + \frac{1}{C_E s} + \frac{1}{C_F s},$$ (5.1)

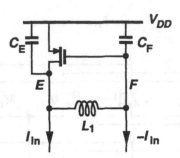

Figure 5.5.   Negative resistance to boost the gain.

where C_E and C_F denote the total capacitance at nodes E and F, respectively. Although in the new topology the conversion gain can be boosted, the circuit is susceptible to oscillation.

To avoid the problems mentioned above, we have used only one of the outputs of the mixer. The circuit is shown in Fig. 5.6. Since only one

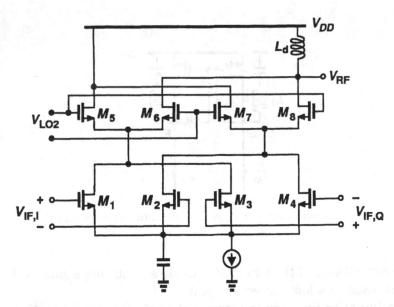

Figure 5.6.   Second upconversion with single-ended output.

of the outputs is used, the conversion gain seems to be 6 dB lower than the differential case (Fig. 5.4). Nevertheless, the parasitic capacitance at the output node is quite lower in this case, leading to a higher inductor value.

While the signal paths from the inputs to the output are symmetric in this circuit, there are more spurs at the output compared to the differential-output case. One of these components is the second harmonic of the LO. To see how this component appears at the output, we can refer to the circuit shown in Fig. 5.7. For a simple analysis, the input signals

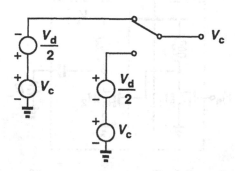

Figure 5.7. Common-mode effect in a single-ended output mixer.

are modeled by differential and common-mode voltages. Ideally, if the switching is abrupt, the common-mode signal goes to the output without any modulation. However, in a practical case where the switching is not perfect, this signal is modulated by narrow pulses on each transition of the LO. Since the common-mode signal is the same for both terminals of the input, the frequency of these narrow pulses is twice that of the LO. In an active mixer, the common-mode signal is the bias current of the mixer, producing the second harmonic of the LO at the output. Nonetheless, the outputs of the circuit and the following stage (PA) are tuned to the RF. Thus, the second harmonic of the LO is small enough before reaching the antenna.

The second upconversion stage draws 1.5 mA from the supply and the RF inductor is realized as a two-layer stacked structure.

4. Power Amplifier

As shown in Fig. 5.1, after two upconversions, the RF signal drives the power amplifier. PAs are typically the most power-hungry building blocks in the transmit path. The PA presented in this section is designed to deliver 1 mW (0 dBm) to the antenna. PAs have been traditionally categorized under many classes: A, B, C, D, E, F, etc. Among these classes, class-A amplifiers are the most linear but the least efficient. Since the output power in our case is not very high, we choose class A amplifier to improve linearity and we use stacking techniques to reduce power consumption.

To achieve enough drive capability, the PA consists of two tapered stages: a driver and an output stage. Figure 5.8 depicts the circuit of a two-stage PA. This circuit, however, suffers from three important

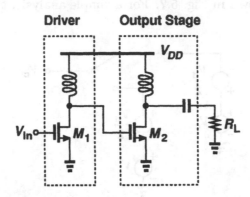

Figure 5.8. Two-stage power amplifier.

drawbacks. First, since there are two current paths from the supply to ground, the power consumption is high. Second, large output swings at the drain of M_2 degrade the long term reliability of the device. Third, in order to deliver 1 mW to a 50 Ω load (R_L), the output stage requires a peak-to-peak current (I_{PP}) of 12.6 mA ($P = I_{PP}^2 R_L/8$). For M_2 to operate as a class A amplifier, the absolute minimum bias current is half of I_{PP}, but to achieve reasonable linearity, the bias current must be higher, leading to high power consumption.

To resolve the first two issues, the circuit is modified using stacking techniques. Figure 5.9 shows the circuit in which the driver is stacked on top of the output stage. Here, bypass capacitor C_b provides ac ground at the source of M_1 and capacitor C_c couples the output of the driver to the input of the output stage. Since this circuit has only one current path from the supply, we can save power. In addition, the drain of M_2 is biased around $V_{DD}/2$, thereby protecting the device from excessive drain-gate voltage.

As mentioned before, if the output of M_2 is directly connected to the load, the circuit demands a large bias current. However, since the required peak-to-peak voltage swing at the output is only 0.63 V (much lower than $V_{DD} = 2.5$ V), a matching network can be employed to increase the output current of M_2. Figure 5.10 shows the matching circuit, which can be viewed as two L-match sections: L_2-C_1 and C_2-L_3. Each L-match section increases the impedance seen from M_2 looking into the load. This network can also be considered as a transformer consisting of C_1 and C_2, which increases the output current of M_2. In

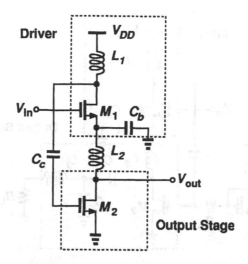

Figure 5.9. Stacking the driver on top of the output stage.

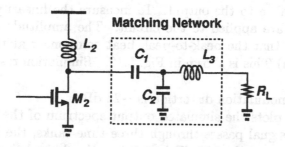

Figure 5.10. Matching network to boost the output current.

this work, C_1 and C_2 are realized on the chip and L_3 is a wirebond inductor.

Figure 5.11 shows the complete PA circuit. In this circuit, a diode connected device (M_b) and a large n-well resistor (R_b) are used to bias the PA. The output inductors (L_1 and L_2) are two-layer stacked spirals. Bypass capacitor C_b is realized by MOS devices and coupling capacitor C_c is a three-layer metal sandwich. The bypass capacitor is designed to have a high quality factor at RF (more than 20) and the connection of the coupling capacitor is such that its bottom-plate capacitance is resonated out by L_1. This is similar to the first downconversion circuit in the receive path. Capacitors C_1 and C_2 are also metal sandwich capacitors.

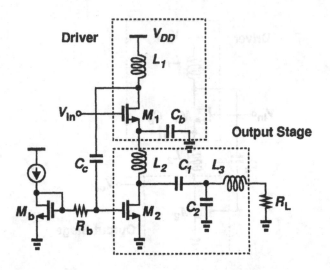

Figure 5.11. Power amplifier circuit.

Simulation results show that with a bias current of 3 mA, the circuit can deliver 0 dBm to the output. To measure the linearity of the PA, two RF-tones are applied to the circuit. The amplitude of the tones is chosen such that the peak-to-peak beat component at the output is 0.63 V (0 dBm) This is shown in Fig. 5.12. Simulation results indicate

that the intermodulation distortion is −28 dBc.

Figure 5.13 plots the simulated output spectrum of the transmitter. Since the RF signal passes through three tune tanks, the image signal at 800 MHz is more than 55 dB lower than the desired signal.

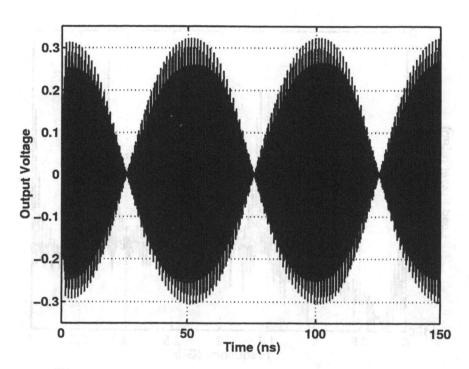

Figure 5.12. Two-tone test to measure intermodulation distortion.

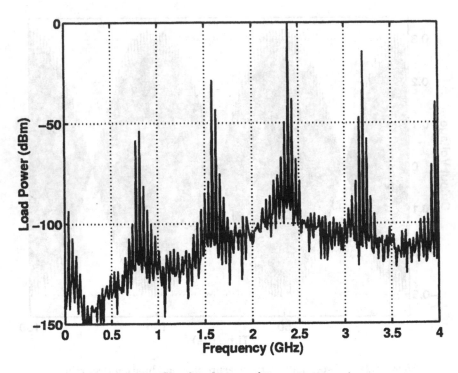

Figure 5.13. Simulated transmitter output spectrum.

Chapter 6

CHANNEL-SELECT FILTER

1. Introduction

After two downconversions in the receive path, the baseband signal must be detected. As explained earlier in Chapter 2, the baseband signals can be demodulated in either the digital or the analog domain. Accompanied by interferers in both cases, the desired signal must be applied to a channel-select filter. For analog detection, the filter must be designed such that the demodulator can detect the signal in the presence of interferers.

In digital demodulation, on the other hand, the baseband signals can be filtered in the digital domain. In this case, however, the analog-to-digital converter (ADC) must be able to handle large interferers, demanding impractically wide dynamic range. Therefore, in order to relax the ADC requirements, the baseband signals are filtered in the analog domain [36, 37].

This chapter describes the channel select filter. We first introduce the concept of noninvasive filtering and then present the filter implementation. Since the channel-select filter needs to be designed for a specific standard, we use Bluetooth specifications to design the filter. Therefore, before describing the filter implementation, we briefly review the baseband signal and interferers in Bluetooth.

2. Noninvasive Filtering

2.1. General Idea

The conventional approach to filtering requires that both the signal and the interferers travel through a circuit that provides the desired transfer function [Fig. 6.1(a)]. However, such a filter introduces significant noise and intermodulation in the signal band. It is therefore advantageous to seek a method that applies filtering to only interferers without invading the signal band. For example, as illustrated in Fig. 6.1(b), a complex impedance $Z_F(s)$ can be placed in parallel with

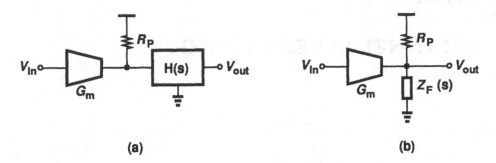

(a)   **(b)**

Figure 6.1.   (a) Conventional filtering (b) noninvasive filtering.

the signal path such that it operates as an open in the signal band while shunting the interferers to ground. As a result, $Z_F(s)$ provides selectivity with negligible additional noise, a critical advantage in view of the high $1/f$ corner frequency in modern CMOS devices. Furthermore, $Z_F(s)$ creates a small intermodulation current through R_P because its Thevenin equivalent is relatively high in the signal band. Nevertheless, some linearity is still necessary if $Z_F(s)$ is to operate as an effective shunt at interferer frequencies.

An example of $Z_F(s)$ for a low-pass filter (LPF) is a simple capacitor but with a capacitor, the order of the filter cannot exceed one. To increase the order of the filter to two (a biquad section), the voltage across the capacitor can be increased as the frequency rises. Figure 6.2 shows a g_m-C implementation of such an impedance. In this circuit, transconductors G_{m2} and G_{m3} form a gyrator that converts capacitor C_L to an emulated inductor L_L at the output of G_{m1}. It can be easily shown that [34]

$$L_L = \frac{C_L}{G_{m2}G_{m3}}. \tag{6.1}$$

Now as the frequency rises, the voltage across C_F increases, giving a second order impedance function. In this implementation, there is a feedforward current path through capacitor C_F that gives a zero in the impedance function. The zero frequency is at

$$\omega_z = \sqrt{\frac{G_{m2}G_{m3}}{C_FC_L}}. \tag{6.2}$$

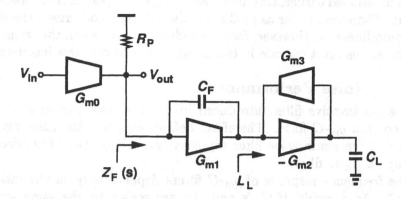

Figure 6.2. Second order noninvasive filter topology.

This is in fact the frequency at which C_F and L_L resonate. To remove this zero (for a Butterworth or type-I Chebyshev filter), a buffer can be added in series with C_F, but to achieve more selectivity , we can exploit the zero as a notch in the adjacent channel. Thus, to realize an elliptic or type-II Chebyshev filter, the circuit in Fig. 6.2 can be used. Writing KVL and KCL equations gives the impedance function of the circuit as

$$Z_F(s) = \frac{1 + \dfrac{C_F C_L}{G_{m2} G_{m3}} s^2}{1 + \dfrac{G_{m1} C_L}{G_{m2} G_{m3}} s} \cdot \frac{1}{C_F s}, \tag{6.3}$$

and the total impedance function is equal to

$$Z_{total}(s) = Z_F(s) \| R_P$$

$$= R_P \frac{1 + \dfrac{C_F C_L}{G_{m2} G_{m3}} s^2}{1 + R_P C_F s + \dfrac{C_F C_L}{G_{m2} G_{m3}} (R_P G_{m1} + 1) s^2}, \tag{6.4}$$

giving the filter transfer function as

$$\frac{V_{out}}{V_{in}}(s) - G_{m0} R_P \frac{1 + \dfrac{C_F C_L}{G_{m2} G_{m3}} s^2}{1 + R_P C_F s + \dfrac{C_F C_L}{G_{m2} G_{m3}} (R_P G_{m1} + 1) s^2}. \tag{0.5}$$

With the proper choice of transconductance and capacitance values, Eq. (6.5) can be used to realize a second order LPF with a pair of zeros (e.g. an elliptic filter).

As mentioned earlier, this filter has negligible impact on the passband signal. Therefore, as far as in-channel distortion is concerned, the filter has good linearity. However, for out-of-channel interferers, the transconductor stages must operate in the linear region to suppress interferers.

2.2.   Noise Performance

The noninvasive filter introduced in the previous section is implemented in a g_m-C form. Therefore, before analyzing the noise performance of the noninvasive filter topology, we briefly study the effect of scaling on g_m-C filters.

The frequency response of g_m-C filters depends only on the ratio of G_m/C. As a result, if G_m's and C's are scaled by the same factor, the transfer function does not change. To see how scaling can affect the noise performance, the noise of each transconductor stage can be modeled by a current source at the output. This is shown in Fig. 6.3. Since all transconductors are composed of the same unit circuit (but

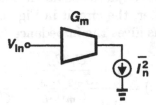

Figure 6.3.   Transconductor noise model.

various number of units), if we neglect the flicker noise, the output noise current is proportional to the transconductance value (G_m) and it can be specified as

$$\overline{i_n^2} = 4kT\Gamma G_m, \tag{6.6}$$

where Γ is a excess noise factor determined by the design of the transconductor stage. For example, for a long-channel MOS transistor, this factor is $2/3$, whereas for a more complicated circuit, it can be quit higher. As Eq. (6.6) shows, the output noise current is proportional to G_m. Therefore, the input-referred noise voltage is inversely proportional to G_m. When G_m and C are increased by the same factor, voltage signals remain the same at all nodes but the input-referred noise voltage of transconductors is reduced, improving the noise performance of the filter. In fact, this improvement is at the expense of the extra power consumed in transconductors (because of higher inductance values) and larger capacitors.

Now that we have seen the effect of scaling on g_m-C filters, we can return to the noninvasive filter. In this circuit, we can model the noise of each transconductor by a current source at the output as shown in Fig. 6.4. The output noise of the filter can be given as

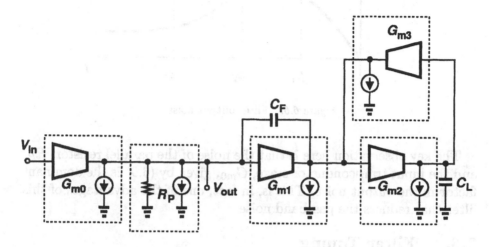

Figure 6.4. Filter noise model.

$$v_{n,out}^2 = \overline{v_{n,G_m}^2} + \overline{v_{n,R_P,G_{m0}}^2}, \tag{6.7}$$

where $\overline{v_{n,G_m}^2}$ is the noise produced by transconductors (G_{m1}-G_{m3}) and $\overline{v_{n,R_P,G_{m0}}^2}$ is the noise

Writing KVL and KCL equations calculates these quantities as

$$\overline{v_{n,G_m}^2} = 4kT\Gamma \cdot \frac{R_P^2 (\frac{C_F C_L}{G_{m2} G_{m3}})^2 \omega^2 [\omega^2 (G_{m1} + G_{m3}) + \frac{G_{m3}^2 G_{m1}}{C_L^2}]}{[1 - \frac{C_F C_L}{G_{m2} G_{m3}}(R_P G_{m1} + 1)^2 \omega^2]^2 + R_P^2 C_F^2 \omega^2}, \tag{6.8}$$

$$\overline{v_{n,R_P,G_{m0}}^2} = 4kT(\frac{1}{R_P} + \Gamma G_{m0})|Z_{total}(j\omega)|^2, \tag{6.9}$$

where $Z_{total}(j\omega)$ is the filter transfer function given by (6.4).

Since Eq. (6.8) has the same poles as $Z_{total}(j\omega)$, the noise of transconductors is shaped by the same poles as the signal. Furthermore, there is a zero at DC that eliminates the noise at low frequencies. Therefore, the effect of flicker noise, which is not considered in this calculation, is quite small. There is also another zero in Eq. (6.8) which is typically higher than the poles. Figure 6.5 plots the output noise of the filter produced by transconductor stages.

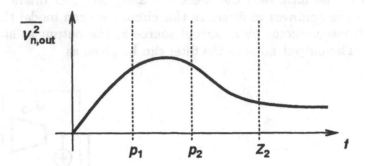

Figure 6.5. Filter output noise.

The key observation here is that the noise of the parallel resistor, R_P, and the input transconductor stage, G_{m0}, given by (6.9) is the dominant factor in the output noise. This is, in fact, one of the advantages of this filter that reduces the passband noise.

2.3.   Filter Tuning

Due to process variations, temperature effects and aging, any active filter needs a dynamic mechanism for tuning the frequency response. In g_m-C filters, all transconductors and capacitors are made of unit cells. The mismatches between similar devices on the same chip is quite small but the absolute values can vary by 20%. Since the transfer function of g_m-C filters depends on the ratio of G_m/C, there must be a mechanism by which this ratio can be corrected.

A number of methods have been introduced to tune g_m-C filters. One of the most popular methods is the VCO tuning loop [34]. In this method, transconductors are designed such that their transconductance can be tuned by a control line. If two transconductor stages followed by capacitors are connected in a loop, a voltage-controlled oscillator (VCO) is made (Fig. 6.6). In this VCO, the frequency of oscillation is

$$\omega_{VCO} = \frac{G_m}{C}. \tag{6.10}$$

Therefore, the frequency of oscillation can be varied by the control line of the transconductors. This VCO is used in a phase-locked loop (PLL) circuit as shown in Fig. 6.6. The external reference of the PLL, coming from a precise crystal oscillator, is chosen so as to set G_m/C (the oscillation frequency of the VCO) to the desired value of the filter. This control line is then used to tune all transconductor stages of the filter.

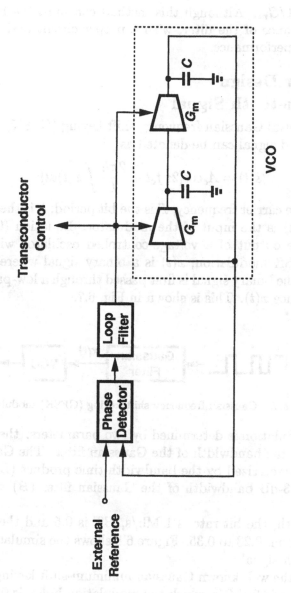

Figure 6.6. VCO tunning loop.

The same approach can be used to tune the noninvasive filter. In the noninvasive filter, however, there is a parallel resistor, R_P, whose variation is not necessarily correlated with that of transconductors. In order to avoid this problem, we can use a transconductor stage to make the resistor. If the output of a transconductor is connected to the input (with proper polarity), the transconductor stage exhibits an equivalent resistance of $1/G_m$. Although this method can limit the linearity and noise performance of the filter, with a proper circuit design, the filter still has high performance.

3.   Filter Design
3.1.   Bluetooth Signal

Bluetooth uses Gaussian frequency shift keying (GFSK) modulation. The modulated signal can be denoted as

$$s(t) = A_c \cos[2\pi f_c t + \frac{2\pi h}{T} \int x(t)dt] \qquad (6.11)$$

where f_c is the carrier frequency, T is the bit period, h is the modulation index and $x(t)$ is the input of the modulator. Equation (6.11) can be viewed as the output of a voltage-controlled oscillator with an input of $x(t)$. In FSK modulation, $x(t)$ is a binary signal whereas in GFSK modulation, the binary signal is first passed through a low-pass Gaussian filter to produce $x(t)$. This is shown in Fig. 6.7.

Figure 6.7.   Gaussian frequency shift keying (GFSK) modulation.

GFSK modulation is determined by two parameters: the modulation index, h, and the bandwidth of the Gaussian filter. The Gaussian filter is usually characterized by the bandwidth-time product (BT), which is equal to the 3-dB bandwidth of the Gaussian filter (B) times the bit period (T).

In Bluetooth, the bit rate is 1 Mb/s, BT is 0.5 and the modulation index varies from 0.28 to 0.35. Figure 6.8 shows the simulated spectrum of a Bluetooth signal.

Note that the well-known Gaussian minimum-shift keying (GMSK) is a special case of GFSK in which the modulation index is 0.5.

3.2.   Single-Tone Interferers

In Bluetooth, the channel spacing is 1 MHz. The interferer in the adjacent channel can be as strong as the desired signal and the alternate channel interferer at 2 MHz can be 30 dB higher than the desired signal. To test the performance of the receiver in the presence of interferers in the adjacent and alternate channels, the desired signal level is set 10 dB above the reference sensitivity (the sensitivity level is −70 dBm).

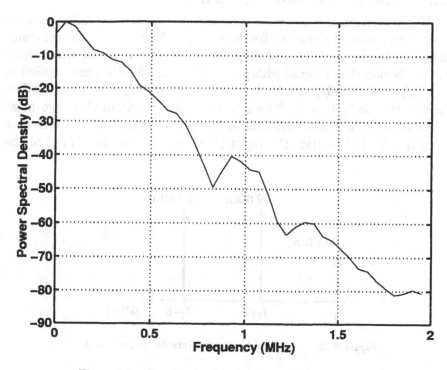

Figure 6.8. Simulated spectrum of a Bluetooth signal.

For channels that are 3 MHz or more farther from the desired signal, the interferer can be 40 dB higher than the desired signal. For these interferers, the desired signal is 3 dB above the reference sensitivity level. Figure 6.9 summarizes the interferer specifications. These interferers do

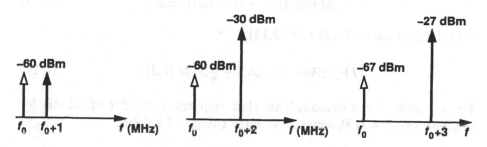

Figure 6.9. Interferers in Bluetooth.

not exist at the same time. Therefore, for two interferers, the intermodulation specification, which is more relaxed, applies.

3.3.   Intermodulation Specification

The intermodulation component is measured under the following conditions: the wanted signal at frequency f_0 is 6 dB above the reference sensitivity level ($-70+6= -64$ dBm). An unmodulated sine wave and a Bluetooth modulated signal with power levels of -39 dBm are applied at f_1 and f_2, respectively, such that $f_0 = 2f_1 - f_2$ and $|f_2 - f_1| = n \times 1$ MHz, where n can be 3, 4, or 5. The bit error rate (BER) in this case must be less than 0.1%. Figure 6.10 shows the signal levels for the worst case where $n = 3$. To calculate the input third intercept point (IIP_3) of the

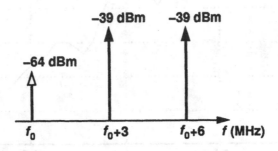

Figure 6.10.   Signal and two-tone interferers for $n = 3$.

system, we can write [1]

$$IIP_3(\text{dBm}) = \frac{1}{2}[-39 - IM(\text{dBm})] - 39, \qquad (6.12)$$

where IM is the input-referred intermodulation level. Note that all signal levels are expressed in dBm in (6.12). The desired signal in this test is -64 dBm. Thus, the signal-to-intermodulation ratio, $SIMR$, is

$$SIMR(\text{dB}) = -64 - IM(\text{dBm}), \qquad (6.13)$$

and if we combine (6.12) and (6.13),

$$IIP_3(\text{dBm}) = -26.5 + \frac{1}{2}SIMR(\text{dB}). \qquad (6.14)$$

For example, for a demodulator that requires a $SIMR$ of 18 dB for BER=0.1%, the IIP_3 must be greater than -17.5 dBm.

3.4.   Order of Filter

As mentioned before, for both digital and analog detection, a channel select filter is required. If the signal is demodulated in the digital domain, the channel select filter is designed to relax the dynamic range of the

ADC, whereas in the case of analog demodulation, the filter is designed such that the demodulator can detect the signal.

If the I and Q signals are specified as

$$I = \cos\phi, \tag{6.15}$$

$$Q = \sin\phi, \tag{6.16}$$

a common solution to detect the signal is

$$
\begin{aligned}
y(t) &= \frac{d}{dt}\tan^{-1}\frac{Q}{I} \\
&= \frac{d}{dt}\tan^{-1}\frac{\sin\phi}{\cos\phi} \\
&= \frac{d\phi}{dt}
\end{aligned}
\tag{6.17}
$$

While this expression suggests high complexity in analog (or digital) implementations, if the differentiation is carried out, we obtain

$$
\begin{aligned}
y(t) &= I\frac{dQ}{dt} - Q\frac{dI}{dt}, \\
&= \frac{d\phi}{dt}\cos^2\phi + \frac{d\phi}{dt}\sin^2\phi, \\
&= \frac{d\phi}{dt}.
\end{aligned}
\tag{6.18}
$$

Figure 6.11 shows the implementation of this demodulator. This is indeed the approach used in the FSK receiver of [40].

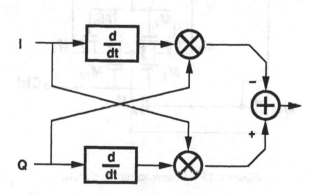

Figure 6.11. GFSK demodulator.

The filter presented here is designed for such an analog demodulator. Simulation results in Matlab indicate that two second-order elliptic filters preceded by a first order LPF sufficiently attenuate interferers, allowing both demodulators mentioned above to detect the desired signal with a BER of less than 0.1%. The filter consists of a first-order LPF with a 3-dB bandwidth of 1 MHz and two second-order elliptic filters with bandwidths of 0.4 and 0.6 MHz, passband ripples of 1 dB, and stopband attenuation of 14 and 29 dB, respectively. Note that two second-order filters exhibit less sensitivity than a fourth-order filter does.

4.   Filter Realization

4.1.   Transconductor Stage

A transconductor is the most important building block in g_m-C filters. Critical parameters in a transconductor design are linearity, noise, tuning range and power consumption.

Figure 6.12 illustrates the transconductor stage. In this circuit, tran-

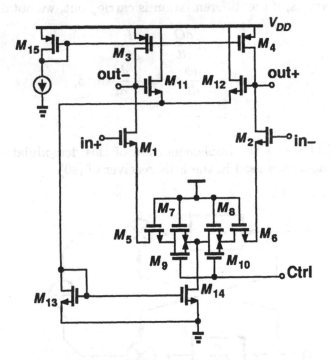

Figure 6.12.   Transconductor circuit.

sistors M_1 and M_2 serve as input devices and transistors M_3 and M_4 as active loads. The channel lengths of these transistors are large enough to ensure high output resistance and low flicker noise (flicker noise is inversely proportional to $W \times L$.). A diode-connected device, M_{15}, biases the load transistors and M_{11}-M_{14} provide common-mode feedback.

To linearize the circuit, degeneration resistors are used in the source of the input devices. The degeneration resistor is a combination of three MOS devices in the triode region, i.e., M_5, M_7 and M_9 on one side and M_6, M_8 and M_{10} on the other side. In this combination, two transistors have constant resistance (M_5,M_7 and M_6,M_8) whereas the third ones (M_9 and M_{10}) are connected to a control line to tune the transconductance value. Connecting the gates of constant resistors to V_{DD} increases their gate-source voltages, making them more linear than variable resistors M_9 and M_{10}. The more the contribution of constant resistance, the more linear the degeneration resistor but this limits the tuning range of the transconductor. Therefore, there is a trade-off between linearity and tuning range of the degeneration resistor. Transistors M_5-M_8 can be replaced by high-resistive poly resistors, but we are limited to a standard digital CMOS process. So, we have used MOS transistors.

Simulation results show that the circuit can provide a transconductance value of 25 $\mu\Omega^{-1}$ (unit cell) with a power consumption of 34 μW. The tuning range of the transconductance is $\pm\%20$ which is shown in Fig. 6.13. The excess noise factor of the stage (defined as (6.6)) is 2.7. Since the circuit is differential, G_m is defined as $(i_{out+} - i_{out-})/(v_{out+} - v_{out-})$ and the excess noise factor is twice that of a similar single-ended circuit.

4.2. Input Transconductor and Parallel Resistor

As mentioned earlier, in order to tune the filter, resistor variations must be correlated with those of transconductors. One simple solution is using a diode connected transconductor which is made by connecting in_+ and in_- of the circuit shown in Fig. 6.12, to out_- and out_+, respectively. In this case, the transconductor circuit exhibits an equivalent resistance of $1/G_m$.

If we look at the filter topology again (Fig. 6.2), we see that resistor R_P is preceded by the input transconductor stage. On the other hand, the PMOS devices in the transconductor circuit only provide bias current with high output resistance. Therefore, we can combine the input transconductor and the diode connected resistor circuit by replacing the PMOS devices with a transconductor stage. The circuit is shown in Fig. 6.14. In this circuit, transistors M_3 and M_4 convert the input voltage signals to current signals. The diode connected circuit then provides the proper resistance value at the output. In order to have more control on noise performance of the input devices, extra current paths are provided by transistors M_{15} and M_{16}.

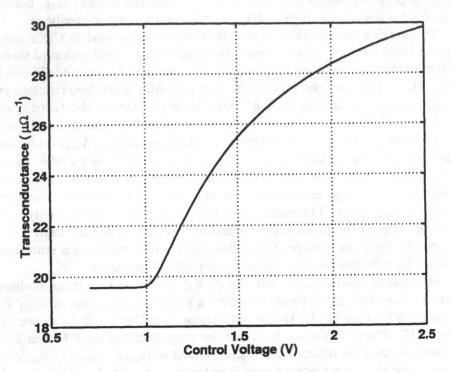

Figure 6.13.   Transconductor tuning range.

Simulation results indicate that the circuit has a voltage gain of 10 dB, with an output resistance of 40 kΩ (single-ended) and consumes 70 μW.

4.3.   Capacitors

If we look at the filter topology (Fig. 6.2), we see that two types of capacitors are used: grounded and floating capacitors, i.e., C_L and C_F, respectively. Since the capacitance values in the LPF can be as high as tens of pF, standard metal capacitors would occupy substantial chip area. Furthermore, they contain considerable bottom-plate capacitance (10% to 20%), which limit their usage as float capacitors. In some technologies, high-value metal-insulator-metal (MIM) capacitors are available. In this design, however, we are restricted to a standard digital CMOS technology. Therefore, we use MOS devices to build capacitors.

For a grounded capacitor we can simply use a MOS transistor as shown in Fig. 6.15(a). Since the output of transconductor stages are biased around the middle of the supply voltage, the MOS device stays in strong inversion. On the contrary, floating capacitors sustain equal voltages across. To ensure strong inversion in these capacitors, we can

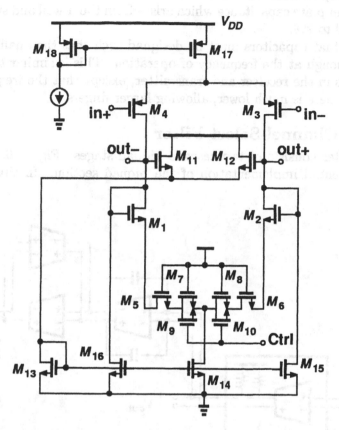

Figure 6.14. Input transconductor and parallel resistor.

used two back-to-back devices as shown in Fig. 6.15(b). Here, the gate

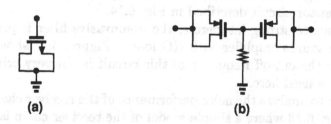

(a)   **(b)**

Figure 6.15. Capacitors in filter: (a) grounded (b) floating.

of the PMOS transistors are biased to ground by a large n-well resistor
(1 MΩ). Using PMOS devices has another advantage as well. If the
source and drain of each device is connected to its well, there will be no

junction capacitance associated with the source and drain. As a result, the bottom-plate capacitance which arises from the n-well and substrate, is reduced to 2%.

Recall that capacitors must be designed such that the quality factor is high enough at the frequency of operation. This is similar to bypass capacitors in the receiver and transmitter, except that the frequency of operation here is much lower, allowing larger fingers.

4.4.   Channel-Select Filter

The filter consists of two cascade biquad stages. Figure 6.16 shows the differential implementation of the biquad section. In this circuit,

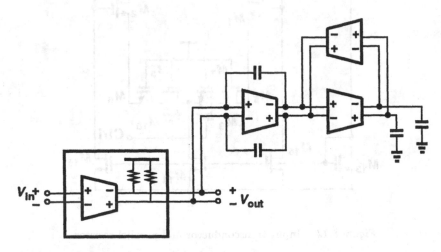

Figure 6.16.   Differential implementation of the filter.

the input stage is the combination of the input transconductor and the parallel resistor circuit described in Fig. 6.14.

To attenuate strong interferers, the noninvasive filter is preceded by a common-source amplifier with RC loads. Figure. 6.17 shows the circuit. Since the cut-off frequency of this circuit is not very critical, poly resistors are used here.

In order to analyze the noise performance of the receiver chain, we can refer to Fig. 6.18 where a simple model of the receiver chain is depicted. If we neglect the noise of noninvasive filers, the noise factor of the receiver can be calculated as

$$F = F_{RF} + \frac{\Gamma}{R_s A_{v,RF}^2/4}\left(\frac{1}{G_{m1}} + \frac{1}{A_{v1}^2 G_{m2}} + \frac{1}{A_{v1}^2 A_{v2}^2 G_{m3}}\right)$$

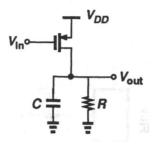

Figure 6.17. Baseband amplifier followed by RC filter.

$$+ \quad \frac{1}{R_s A_{v,RF}^2 / 4}\left(\frac{R_1}{A_{v1}^2} + \frac{R_2}{A_{v1}^2 A_{v2}^2} + \frac{R_3}{A_{v1}^2 A_{v2}^2 A_{v3}^2}\right) \qquad (6.19)$$

where F_{RF} and $A_{v,RF}$ are the noise factor and voltage gain of the RF front end, respectively, Γ is the excess noise factor of transconductors and $A_{v1} - A_{v3}$ are the voltage gain of the baseband stages ($A_{vi} = G_{mi}R_i$, i=1,2,3). After careful analysis, we designed the filter for a total noise figure of 5.5 dB, which is far below the required value for Bluetooth. The power consumption of the I and Q filters is 4 mW. Figure 6.19 shows the simulated characteristic of the filter and Fig. 6.20 plots the output noise of the filter (excluding the noise of the RF front end and RC amplifiers). Since we are using long-channel devices, the default value of 2/3 is used for the excess noise factor of transistors. Simulation results also indicate that the output noise level of the input transconductor and the parallel resistor is 147 nVHz$^{-1/2}$. This value is equal to the dc output noise of the biquad sections and it reveals that the noise contribution of the nonintrusive filter is much lower than that of the input transconductor and the parallel resistor. Table 6.1 summarizes the transconductors and capacitors values of the two biquad sections.

	First Section	Second Section
G_{m1} (mΩ^{-1})	0.35	0.6
G_{m2} (mΩ^{-1})	0.1	0.05
G_{m3} (mΩ^{-1})	0.1	0.2
R_P (kΩ)	10	40
C_F (pF)	31.7	6.35
C_L (pF)	9.38	3.9

Table 6.1. Component values for the filter.

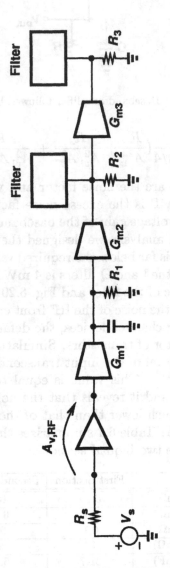

Figure 6.18.   Receiver chain.

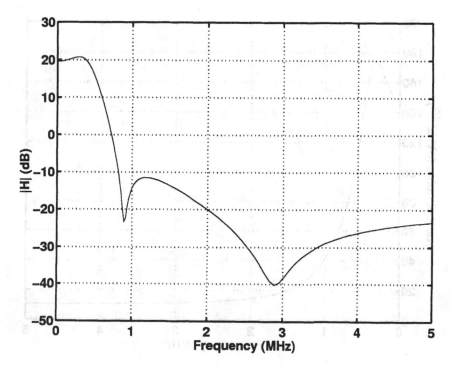

Figure 6.19. Simulated characteristic of the biquad sections.

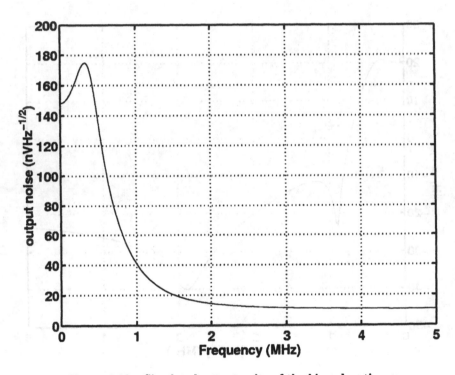

Figure 6.20. Simulated output noise of the biquad sections.

Chapter 7

EXPERIMENTAL RESULTS

1. Introduction

The final phase of this research culminates in the design and fabrication of two prototypes. The first is a transmitter/receiver front end operating in the 2.4 GHz ISM band. The second uses the same RF front end followed by a noninvasive channel select filter in the receive path. Since the frequency planning of transceiver can accommodate a GPS receive path, a GPS receiver front-end is also included in the second prototype. This chapter describes the test setup and presents the experimental results of the two prototypes.

2. Test Setup
2.1. Equipment

Since the transceiver does not include the frequency synthesizer, the device under test (DUT) requires a number of RF signal generators for local oscillators and the RF input. These RF signals are provided by the HP E4422B signal generators and a dual-channel low-frequency HP 3326A signal generator for the transmitter test. Since RF generators do not provide differential signals, (required in LO's), Mini-Circuit and Anaren power splitters are used.

The input return loss (S_{11}) of the receiver is measured with the HP 8720D network analyzer. Frequency-domain data is mainly analyzed with the Rhode and Schwartz FSEB spectrum analyzer but HP 8562A, and HP 3585A (for low frequencies) were also used. The noise figure of the receiver is measured with the HP 8970B noise figure meter and the two noise sources: HP 346A with an ENR of 5 dB and Noisecom NC346B with an ENR of 15 dB.

2.2.   Test Board

In order to avoid the parasitics associated with the package, a chip-on-board assembly is used. The test board is made from a 31-mil double-sided copper-clad board with an FR-4 dielectric. Transmission lines having a 50-Ω impedance are designed on the top layer of the board. Figure 7.1 shows the test board layouts used in the transceiver.

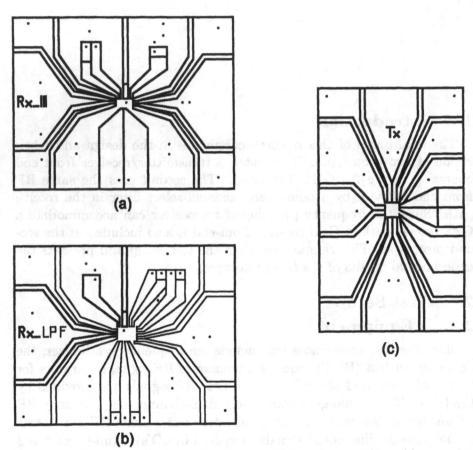

Figure 7.1.   Board layouts (a) receiver front end (b) receiver and filters (c) transmitter.

2.3.   Noise Figure Measurements at Low Frequencies

The minimum frequency of the noise figure meter is 10 MHz. Therefore, in order to measure the noise of the overall receiver chain containing low-pass filters (LPF), a spectrum analyzer must be used. A simple method to measure the noise figure is to obtain the output noise floor of the receiver. In this method, the measured noise power can be specified as

$$N_o = (kT + N_{a,in})GB \tag{7.1}$$

where $N_{a,in}$ is the amplifier noise referred to the input, G is the gain and B is the resolution bandwidth of the spectrum analyzer. Then, the noise factor can be calculated as

$$F = 1 + \frac{N_{a,in}}{kT} \tag{7.2}$$

The main disadvantage of this method is that any error in N_o, B or G (as (7.1) shows) degrades the accuracy.

To alleviate these issues, we apply the hot-cold method, which is used by noise figure meters. In this method, the output noise power is measured at two points: the cold point with a source temperature of T_c and the hot point with a source temperature of T_h. Plotting the output noise power as a function of the source temperature (Fig. 7.2) yields

$$N_c = kGB(T_c + T_n), \tag{7.3}$$
$$N_h = kGB(T_h + T_n), \tag{7.4}$$

where N_h and N_c are the measured output noise power at hot and cold

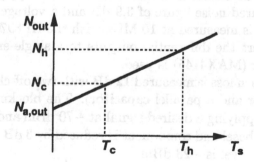

Figure 7.2. Output noise power as a function of source temperature.

points and T_n is the noise temperature of the circuit. The quantity T_n is defined as the temperature at which the source impedance must be held so that the noise output generated by the source impedance is equal to the output noise of the circuit [41]. In other words, the output noise power of the circuit denoted by $N_{a,out}$ is given by

$$N_{a,out} = GN_{a,in}$$
$$= kGBT_n. \tag{7.5}$$

If we divide (7.4) by (7.3),

$$\frac{N_h}{N_c} = \frac{T_h + T_n}{T_c + T_n},$$ (7.6)

the noise temperature of the circuit can be calculated as

$$T_n = \frac{T_h - T_c \dfrac{N_h}{N_c}}{\dfrac{N_h}{N_c} - 1}.$$ (7.7)

This gives the noise factor as

$$F = 1 + \frac{T_n}{T}.$$ (7.8)

where T is the circuit temperature at which the noise performance is specified (usually T_c is the same as T).

As (7.6)-(7.8) show, the noise factor in this method is independent of B and G, requiring only the ratio of N_h/N_c. This relaxes the calibration requirements of the spectrum analyzer.

3. First Prototype: RF Front-End Chip

The RF front end has been fabricated in TSMC 0.25-μm CMOS technology. Figure 7.3 shows the die photograph. The chip area is 1.2×1.1 mm². The circuit is tested with a 2.5-V supply. The receiver achieves a measured noise figure of 3.9 dB and a voltage gain of 40 dB. The noise figure is measured at 10 MHz with the HP8970B noise-figure meter. To convert the differential outputs to a single-ended signal, an off-chip amplifier (MAX4145) is used.

The input return loss is measured 12 dB with two off-chip components (a series inductor and a parallel capacitor). The blocking performance is measured by applying a desired signal at −70 dBm and increasing the blocker until the baseband component is reduced by 3 dB. The maximum tolerable blocker level is −28 dBm.

The input IP_3 is −17.5 dBm, which is limited by baseband amplifiers. Since the noise figure of the receiver is quite lower than the values required by Bluetooth and IEEE802.11b, we can trade noise figure for linearity. In fact, our measurements indicate that if the gate voltage of cascode device in the LNA is lowered, the IIP_3 and noise figure rise at roughly the same rate. The measured image rejection is 41 dB. Table 7.1 summarizes measurement results.

The LNA and mixers consume 6.25 mW and the total power consumption of the receiver is 21 mW. This includes the power hungry baseband amplifiers that are optimized for noise figure measurements.

Figure 7.3.   Receiver front end die photograph.

Noise Figure	3.9 dB
$|S_{11}|$	-12 dB
Image Rejection	41 dB
Voltage Gain	40 dB
Blocking	-28 dBm
IIP_3	-17.5 dBm
Power Dissipation:	
LNA and Mixers	6.25 mW
Divider	4 mW
Baseband Amplifiers	7.25 mW
Total	17.5 mW

Table 7.1.   Measured receiver front end performance.

The transmitter delivers 0 dBm to a 50-Ω load with sideband spurs of less than -30 dBm and it consumes 12 mW from a 2.5 V supply. Table 7.2 summarized the measured performance of the transmitter and Fig. 7.4 and 7.5 show the output spectrum of the transmitter in narrow and wide spans.

Output Power	0 dBm
Sidebands	-30 dBc
Power Dissipation	12mW

Table 7.2. Measured transmitter performance.

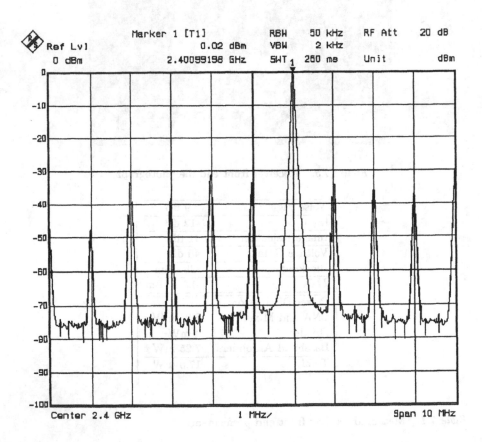

Figure 7.4. Measured transmitter spectrum in a narrow span.

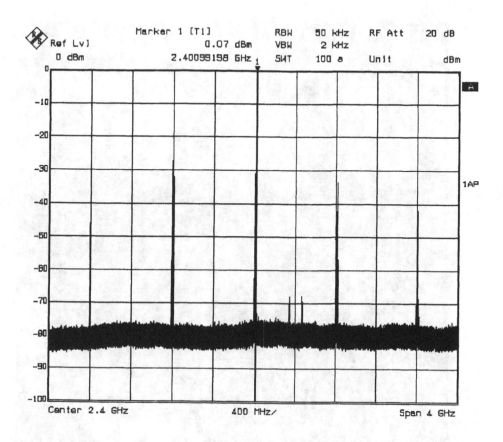

Figure 7.5. Measured transmitter spectrum in a wide span.

4. Second Prototype: Transmitter/Receiver Chip

Figure 7.6 shows the die photograph of the transmitter/receiver fabricated in TSMC 0.25-μm CMOS technology. The chip also includes a GPS receiver front end and it occupies an area of 1.83×2 mm^2.

The 2.4-GHz front end of this chip is the same as the RF front end presented in the previous section. The baseband amplifiers in this prototype are modified for lower power consumption and slightly better linearity. These amplifiers are then followed by noninvasive channel-select filters designed based on Bluetooth specifications.

Since the output of receiver passes through the low-pass filters, the noise figure cannot be measured by HP8970B. Therefore, the hot-cold

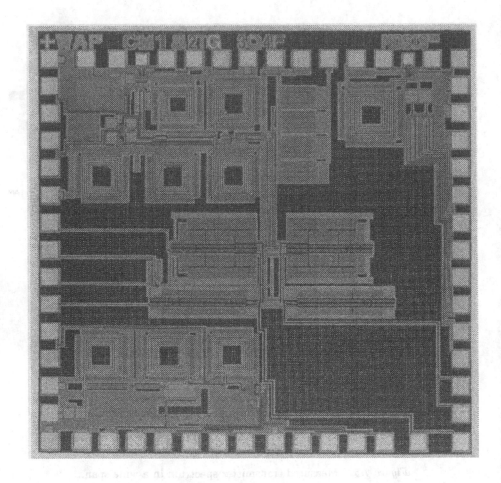

Figure 7.6. Transmitter/receiver die photograph.

method is employed. The receiver achieves a noise figure of 6 dB at 200 kHz with a total power consumption of 17.5 mW from a 2.5 V supply. Table 7.3 summarized the receiver measured performance. Among prior works, [2], realized in a 0.35-μm, consumes the lowest power, reporting 96 mW in the receiver (including the demodulator) with a noise figure of 12 dB.

Figure 7.7 plots the measured characteristic of the receiver. Since the output impedance of the last biquad filter is 40 kΩ, parasitic capacitance

Noise Figure	6 dB
$|S_{11}|$	-12 dB
Image Rejection	41 dB
Voltage Gain	50 dB
$SIMR$	26 dB
Power Dissipation:	
LNA and Mixers	6.25 mW
Divider	3.75 mW
Baseband Amplifiers	3.5 mW
I and Q LPF's	4 mW
Total	17.5 mW

Table 7.3. Measured receiver performance.

can distort the filter characteristic. In order to resolve this issue, off-chip common-collector buffers (NE68719) are employed.

Owing to the compatibility of the LO frequency with the GPS L_1 signal, a GPS front end is also accommodated in this chip. The GPS front end, consisting of an LNA capacitively coupled to I and Q low-IF active mixers, achieves a noise figure of 3.5 dB with a power consumption of 6.25 mW. Table 7.4 shows the measured performance summary of the GPS front end.

Noise Figure	3.5 dB
Gain	36 dB
$|S_{11}|$	-14 dB
IIP_3	-22 dBm
Power Dissipation	6.25mW

Table 7.4. Measured performance of GPS front end.

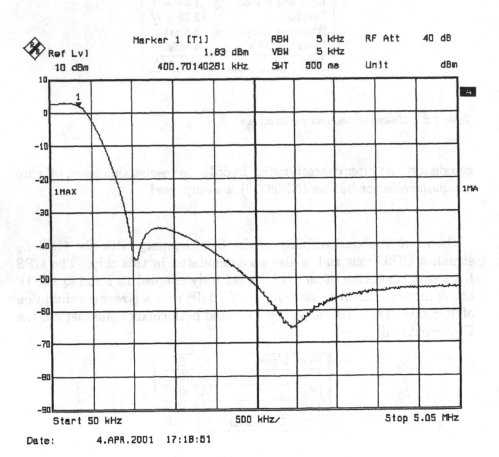

Figure 7.7.   Measured receiver characteristic.

Chapter 8

CONCLUSION

Over recent years, the market for wireless communications has grown enormously. In addition to familiar wireless products like pagers, and cellular phones, other markets such as wireless LANs and PANs display a great potential for rapid growth. GPS receivers have also proven useful for positioning.

This increasing demand has motivated extensive research on low-power solutions. Furthermore, using the CMOS technology with high levels of integration can substantially lower the cost of RF transceivers. In this book, we described architecture, circuits, and passive components to design low-power wireless transceivers in a standard digital CMOS technology.

The transceiver architecture is one of the most challenging aspects of the design, impacting complexity, cost, power dissipation, and the number of external components. While homodyne systems suffer from many issues, heterodyne topologies require image-reject filters, adding more complexity to the system and/or requiring off-chip components. With the proposed architecture, however, we benefit the advantages of the heterodyne architecture, while no explicit image-reject filter is needed. Besides, only one frequency synthesizer is employed for both frequency conversions.

The RF front end is a critical building block in the receiver, which greatly affects the sensitivity and linearity of the system. To provide good performance while achieving low power consumption, we introduced stacking technique. The stacking allows reusing the current of one stage in another, saving substantial power. This technique is also applicable to other RF blocks such as the transmitter. To further reduce the power consumption, we also used passive mixers in the receiver and transmitter. Using passive mixers improves linearity as well.

Analog filters used in the baseband section of RF receivers must satisfy stringent noise, linearity, power dissipation, and selectivity requirements. The existence of large interferers near the desired signal frequency demands a high linearity and/or low noise in the filters, impacting the distribution of gain and noise through the receiver chain. In order to relax the trade-offs between the above parameters, we introduced a non-invasive filtering technique. In this topology, as the name implies, the desired signal is not invaded by the filter. The merits of this topology was demonstrated in a g_m-C implementation. To provide compatibility with the standard digital CMOS process, MOS devices were used to implement all passive components.

Monolithic inductors have found extensive usage in RF circuits to provide a high level of integration. Despite their relatively low quality factor, Q, such inductors still prove useful in providing gain with minimal voltage headroom and operating as resonators in oscillators. Among different structures, stacked structures can provide large inductance values while occupying minimal chip area. With availability of five layers of metal in modern CMOS technology, the choice of the metal layers becomes important. To choose the metal layers for better performance, we have derived a closed form expression, predicting the self-resonance frequency of the inductor with an error of less than of 5%. This expression helps us to improve the self-resonance frequency by 100%.

Monolithic transformers have also appeared in CMOS technology, allowing new circuit configurations. We also fabricated stacked transformers in CMOS technology. Although the transformers were not used in our circuits, measurement results indicate that we can achieve voltage gains of 1.8 and 3 for 1:2 and 1:4 transformers, respectively.

The low-power techniques, architecture, and monolithic inductors were employed to implement a 2.4-GHz transceiver, targeting some of the challenging specifications of the two popular standards in this band, i.e., Bluetooth and IEEE802.11b. Fabricated in a 0.25-μm CMOS technology, the chip consumes 17.5 mW from a 2.5-V supply and achieves a noise figure of 6 dB of which 3.9 dB is due to the RF front end. The transmitter delivers 0 dBm with sidebands of less than -30 dBm. This prototype also accommodates a low-IF GPS receiver.

This research has opened up several possibilities for future research. One important area is the possibility of using stacking techniques for circuits with lower supply voltages. Another area involves the study of stacked spirals for differential inductors. Using stacked transformers with voltage or current gain in RF circuits is also an area of practical importance. Finally, the noninvasive filtering technique, which was implemented as a second-order g_m-C stage, can be investigated for other implementations and/or higher orders.

References

[1] B. Razavi, *RF Microelectronics*, NJ: Prentice Hall, Upper Saddle River, 1998.

[2] H. Darabi *et al.*, "A 2.4GHz CMOS Transceiver for Bluetooth," *ISSCC Dig. of Tech. Papers*, pp. 200-201, Feb. 2001.

[3] F. Op't Eynde *et al.*, "A Fully-Integrated Single-Chip SOC for Bluetooth," *ISSCC Dig. of Tech. Papers*, pp. 196-197, Feb. 2001.

[4] A. Ajjkuttira *et al.*, "A Fully-Integrated CMOS RFIC for Bluetooth Applications," *ISSCC Dig. of Tech. Papers*, pp. 198-199, Feb. 2001.

[5] P. Stroet *et al.*, "A Zero-IF Single-Chip Transceiver for up to 22 Mb/s QPSK 802.11b Wireless LAN," *ISSCC Dig. of Tech. Papers*, pp. 204-205, Feb. 2001.

[6] B. Razavi, "A 2.4-GHz CMOS Receiver for IEEE 802.11 Wireless LAN's," *IEEE Journal of Solid-State Circuits*, vol. 34, No. 10, pp. 1382-1385, Oct. 1999.

[7] H. Darabi and A. Abidi, "Noise in RF Mixers: A Simple Physical Model," *IEEE Journal of Solid-State Circuit*, vol. 35, no. 1, pp. 15-25, Jan. 2000.

[8] J. J. Zhou and D. J. Allstot, "Monolithic Transformers and Their Application in a Differential CMOS RF Low-Noise Amplifier" *IEEE J. Solid-State Circuits*, vol. 33, pp. 2020-2027, Dec. 1998.

[9] C. P. Yue and S. S. Wong, "On-Chip Spiral Inductors with Patterned Ground Shields for Si-Based RF IC's," *IEEE J. Solid-State Circuits*, vol. 33, pp. 743-752, May 1998.

[10] M.W. Green *et al.*, "Miniature Multilayer Spiral Inductors for GaAs MMIC's," *GaAs IC Symposium*, pp. 303-306, 1989.

[11] R. B. Merril *et al.*, "Optimization of High Q Inductors for Multi-Level Metal CMOS," *Proc. IEDM*, pp. 38.7.1-38.7.4, Dec. 1995.

[12] J. Crols *et al.*, "An Analytical Model of Planer Inductors on Lowly Doped Silicon Substrates for High Frequency Analog Design up to 3 GHz," *Dig. of VLSI Circuits Symposium*, pp. 28-29, June 1996.

[13] S. Mohan et al., "Simple Accurate Expressions for Planar Spiral Inductors," *IEEE J. Solid-State Circuits*, vol. 34, pp. 1419-1424, Oct. 1999.

[14] H. M. Greenhouse, "Design of Planar Rectangular Microelectronics Inductors," *IEEE Trans. Parts, Hybrids, Packaging*, vol. PHP-10, pp. 101-109, June 1974.

[15] A. M. Niknejad and R. G. Meyer, "Analysis, Design, and Optimization of Spiral Inductors and Transformers for Si RF IC's," *IEEE J. Solid-State Circuits*, vol. 33, pp. 1470-1481, Oct. 1998.

[16] B. Razavi, "CMOS Technology Characterization for Analog and RF Design," *IEEE J. Solid-State Circuits*, vol. 34, pp. 268-276, March 1999.

[17] J. N. Burghartz et al., "RF Circuit Design Aspects of Spiral Inductors on Silicon," *IEEE J. Solid-State Circuits*, vol. 33, pp. 2028-2034, Dec. 1998.

[18] J. R. Long and M. A. Copeland, "The Modeling, Characterization, and Design of Monolithic Inductors for Silicon RF IC's," *IEEE J. Solid-State Circuits*, vol. 32, pp. 357-369, March 1997.

[19] W. B. Kuhn and N. K. Yanduru, "Spiral Inductor Substrate Loss Modeling in Silicon RF IC's," *Microwave Journal*, pp. 66-81, March 1999.

[20] D. K. Shaeffer, T. H. Lee, "A 1.5-V, 1.5-GHz CMOS Low Noise Amplifier," *IEEE Journal of Solid-State Circuits*, vol. 32, pp. 745-759, May 1997.

[21] T. H. Lee, "The Design of Narrowband CMOS RF Low-Noise Amplifiers," *Advances in Analog Circuit Design*, Copenhagen, Denmark, April 28-30, 1998.

[22] B. Razavi, R. H. Yan, and K. F. Lee, "Impact of Distributed Gate Resistance on the Performance of MOS Devices," *IEEE Trans. Circuits Syst. I*, vol. 41, pp.750-754, Nov. 1994.

[23] H. Darabi, "An Ultralow Power Single-Chip CMOS 900 MHz Receiver for Wireless Paging," Ph.D. dissertation in Electrical Engineering, UCLA, 1999.

[24] L. Der, "A 2-GHz CMOS Image-Reject Receiver with Sign-Sign LMS Calibration," Ph.D. dissertation in Electrical Engineering, UCLA, 2001.

[25] B. Razavi "A 900-MHz/1.8-GHz CMOS Transmitter for Dual-Band Applications," *IEEE J. Solid-State Circuits*, vol. 34, pp.573-579, May. 1999.

[26] J. Crols and M. S. J. Steyaert, *CMOS Wireless Transceiver Design*, Kluwer Academic Publishers, 1997.

[27] F. Behbahani, et al., "An Adaptive 2.4-GHz Low-IF Receiver in 0.6-μm CMOS for Wideband Wireless LAN," *ISSCC Dig. of Tech. Papers*, pp. 146-147, Feb. 2000.

[28] S. Wu and B. Razavi, "A 900-MHz/1.8-GHz CMOS Receiver for Dual-Band Applications," *IEEE Journal of Solid-State Circuits*, vol. 33, pp. 2178-2185, Dec. 1998.

[29] R. Montemayor and B. Razavi, "A Self-Calibrating 900-MHz CMOS Image-Reject Receiver," *Proc. of ESSCIRC*, Sweden, Sep. 2000.

[30] L. Der and B. Razavi, "A 2GHz CMOS Image-Reject Receiver with Sign-Sign LMS Calibration," *ISSCC Dig. of Tech. Papers*, pp. 294-295, Feb. 2001.

[31] B. Razavi, "A 5.2-GHz CMOS Receiver with 62-dB Image Rejection," *IEEE Journal of Solid-State Circuits*, vol. 36, No. 5, pp. 810-815, May. 2001.

[32] A. Chan, "A 1.6-GHz Frequency Synthesizer in 0.25-μm CMOS Technology," M.S. thesis in Electrical Engineering, UCLA, 2000.

[33] R. S. Carson, *High-Frequency Amplifiers*, John Wiley & Sons, 1982.

[34] B. Nauta, *Analog CMOS Filters For very High Frequencies*, Kluwer Academic Publishers, 1993.

[35] R. Schaumann, M. Ghausi and K. Laker, *Design of Analog Filters*, Prentice-Hall, 1990.

[36] B. Razavi, "CMOS RF Receiver Design for Wireless LAN Applications," *IEEE Radio and Wireless Conference, 1999. RAWCON 99*, pp. 275 -280, 1999.

[37] B. Razavi, "Architectures and Circuits for RF CMOS Receivers," *IEEE Custom Integrated Circuits Conference*, pp. 393 -400, May 1999.

[38] Bluetooth Specification, Version 1.0B, Nov. 1999.

[39] IEEE Std 802.11b Draft Supplement to Standard for Information Technology Telecommunications and Information Exchange between Systems Local and Metropolitan Area Networks Specific Requirements, 1999.

[40] J. Wilson *et al.*, "A Single-Chip VHF and UHF Receiver for Radio Paging," *IEEE Journal of Solid-State Circuits*, vol 26, pp. 1944-1950, Dec. 1991.

[41] P. Gray and R. Meyer, *Analysis and Design of Analog Integrated Circuits*, John Wiley & Sons, 1984.

[42] B. Razavi, *Design of Analog CMOS Integrated Circuits*, McGraw-Hill, 2001.

[43] S. Haykin, *Communication Systems*, John Wiley & Sons, 2001.

[44] J. Spilker, "GPS Signal Structure and Theoretical Performance," in B. Parkinson and J. Spilker, Jr., Eds., Global Positioning System: Theory and Applications, vol. I. American Institute of Aeronautics and Astronautics, 1996.

[45] A. Van Dierendonck, "GPS receivers," in B. Parkinson and J. Spilker, Jr., Eds., Global Positioning System: Theory and Applications, vol. I. American Institute of Aeronautics and Astronautics, 1996.

[46] M. Braasch and A. Van Dierendonch, "GPS Receiver Architectures and Measurements," *Proceedings of the IEEE*, vol. 87, pp. 48-64, Jan. 1999.

[47] D. Shaeffer *et al.*, "A 115-mW, 0.5-mm CMOS GPS Receiver with Wide Dynamic-Range Active Filters," *IEEE Journal of Solid-State Circuits*, vol 33, pp. 2219-2231, Dec. 1998

[48] B. Parkinson and J. Spilker, Jr., Eds., *Global Positioning System: Theory and Applications*, vol. I. American Institute of Aeronautics and Astronautics, 1996.

Index